ELEMENTARY ALGEBRA

A Prerequisite for Functions

Preliminary Edition

Darrell H. Abney
Maysville Community College

Kathy R. Mowers
Owensboro Community College

Dana T. Calland
Maysville Community College

Lillie R. F. Crowley
Lexington Community College

▲ ADDISON-WESLEY

An imprint of Addison Wesley Longman, Inc.

Reading, Massachusetts • Menlo Park, California • New York
Harlow, England • Don Mills, Ontario • Sydney • Mexico City
Madrid • Amsterdam

Publisher: Jason Jordan
Senior Editorial Project Manager: Kari Heen
Editorial Assistant: Palmer MacKay
Managing Editor: Ron Hampton
Production Assistant: Jane Estrella
Design Direction: Susan Carsten
Cover Designer: Renée Sartell
Prepress Services Buyer: Caroline Fell
Senior Marketing Manager: Craig Bleyer
Senior Marketing Coordinator: Mark Harrington
Manufacturing Coordinator: Evelyn Beaton

This text was produced using Ami Pro 3.1™ by Lotus Development Corporation.

About the Cover

The lizard graphic associated with the NKATE project was derived from the Tohono O'Odham (Papago) symbol for "change and the courage it takes." The cover concept, created for the Class Test Edition by Kathy Mowers and Steven Paulo Davis, with electronic design and execution by Holly Bowyer, was recreated for the Preliminary Edition by Susan Carsten and Renée Sartell.

Reproduced by Addison Wesley Longman from camera-ready copy supplied by the authors.

Copyright © 1999 Addison Wesley Longman, Inc.
All rights reserved. No part of this publication may be reproduced, stored in a retrieval system, or transmitted, in any form or by any means, electronic, mechanical, photocopying, recording, or otherwise, without the prior written permission of the publisher. Printed in the United States of America.

ISBN 0-201-35198-6

2 3 4 5 6 7 8 9 10   CRS   01 00 99

Contents

We introduce the concept of a variable and use formula evaluation to review arithmetic of decimals, fractions, and integers. Hands-on activities help students learn to work in groups. Students complete input-output tables from given formulas and graphs. They construct tables from formulas and plot points on a rectangular-coordinate system. Students use the graphing calculator to evaluate formulas and to plot points.

Students solve linear equations numerically using input-output and calculator tables. They solve linear equations graphically using paper-and-pencil and calculator graphs. Students solve linear equations algebraically and verify the solutions graphically or numerically.

We use area and perimeter to introduce simplifying polynomial expressions. Students study properties of exponents and scientific notation. Students use estimation skills to verify if their answers are reasonable.

Students review linear equations by solving applications with fractional equations, proportions, direct variation, percent equations, and literal equations. Students solve inequalities, write the solutions in interval and inequality notation, and graph them on the number line.

Unit 5—Slope and Equations of Lines 175

We link the concept of a line and its slope to rate. Students use rise-run triangles and the formula to determine the slope of a line. They use the slope-intercept method to graph lines. When given a line, they write its equation using the slope and y-intercept. The plotting and graphing features of the calculator reinforce the students' understanding of these key concepts.

Unit 6—Factoring and Quadratic Equations 211

Students review multiplication of polynomials and use those skills to develop an understanding of factors, factoring, and factored form. Students find the x-intercepts of quadratic equations algebraically and graphically. Students explore the relationship between the factors, x-intercepts, and zeros of a quadratic. Students simplify algebraic fractions and solve fractional equations. The graphing calculator is used to approximate x-intercepts.

Unit 7—Systems of Linear Equations 253

Students find solutions of linear systems graphically, numerically from tables, and algebraically by the methods of substitution. Solutions of systems without unique solutions lead to a discussion of inconsistent systems and dependent equations. We use the trace and table features of the graphing calculator to approximate the solution to a linear system of equations.

Unit 8—Radicals and Radical Equations 277

Students explore the relationship between perfect squares and square roots. They simplify radical expressions and evaluate expressions with radical and fractional exponents. Students solve radical equations algebraically, graphically, and numerically. Applications of radicals include finding the distance between two points, the velocity of a satellite orbiting the earth, and the period of a pendulum. Students find the area of triangles using Heron's formula.

Mathematical Contents

Preface for the Student

A Different Kind of Text

This text is different from most mathematics textbooks you have used. Experiments and hands-on projects are included to involve you actively in learning mathematics. Examples and application problems from real-world situations are incorporated to help answer the age-old question, "Why do I have to study this?"

The symbols explained below are used throughout the text.

 The lizard icon marks a problem or activity designed to be worked in groups of 3 or 4 students in a collaborative learning environment. Your instructor may ask you to turn in a report based on the activity.

 The pencil-and-paper icon marks areas for student interaction that may not be otherwise obvious.

 The magnifying-glass icon indicates examples you should examine carefully.

Graphing Calculator Used throughout the Course

Calculator technology is used wherever appropriate to increase understanding, to reinforce topics, and to replace tedious manipulations. You will use the graphing calculator to plot points, to solve equations, and to examine relationships numerically. You may find the use of the calculator frustrating at first, but you will develop important calculator skills that will be useful in this and subsequent courses.

Student Responsibility for Learning

You will be actively involved in learning mathematics through activities and collaborative learning. This means that you must come to class with your homework prepared and be ready to contribute to class and group discussions. Don't be afraid to try. You may not be able to work all problems at first, but you will make progress throughout the term. Most units include review problems so you can continue to practice newly learned skills. To succeed in this course, it is critical that you study for understanding rather than memorize steps.

We hope that you enjoy the course as much as the students who used earlier editions. The course will be demanding but far more rewarding than a traditional course.

Prerequisite Skills

To study *Elementary Algebra: A Prerequisite for Functions* you should be able to:

1. Add, subtract, multiply, and divide decimals;

2. Round numbers to a given accuracy;

3. Add, subtract, multiply, and divide fractions;

4. Change decimals and fractions to percents;

5. Change percents to decimals and fractions;

6. Solve basic percent problems;

7. Simplify absolute value expressions;

8. Add, subtract, multiply, and divide integers; and

9. Evaluate expressions using the order of operations agreement.

Preface for the Instructor

Vision

This preliminary edition is the first title in a series of developmental mathematics textbooks developed with a strong student focus while incorporating the American Mathematical Association of Two-Year Colleges (AMATYC) standards for problem solving, modeling, connecting with other disciplines, and technology. Course materials use collaborative activities to engage students.

Elementary Algebra: A Prerequisite for Functions contains the topics the authors believe are needed for further study of algebra, geometry, and technical mathematics. We encourage you and your students to refer to the *Preface for the Student*. Students use the graphing calculator for numerical and graphical explorations. Throughout the text, problems from geometry and the real world show students how mathematics applies to life outside the classroom.

Need for Algebra Reform

Perhaps the biggest problem facing college mathematics programs across the nation is the number of entering students who test into developmental mathematics courses instead of college-level mathematics. This is especially serious in two-year colleges, where more than one-half of entering students begin with a developmental algebra or a prealgebra course. High withdrawal rates in these courses compound the problem. Success rates of 50% or less mean few students survive to study further mathematics, science, or technology. Developmental mathematics courses are generally taught in a very traditional lecture format with emphasis on drills and skills. Uri Treisman has said that in our algebra courses we are saying the same things students have heard (and not understood) for years; we're just saying them faster and louder, and hoping that they will "catch it" this time.

Our desire to change this situation led to the development of this text. The lizard on the cover symbolizes our desire to change the way we teach. We have developed a text with essentially the same goals as the traditional elementary algebra book, but with a different approach and delivery system. We approach each topic graphically, numerically, and algebraically, mindful of the AMATYC standards. We expect students to be able to communicate the mathematics they learn and expect this approach to better prepare them for further study in mathematics, science, and technology.

As William Butler Yeats said, "Education is not the filling of the pail; it is the lighting of the fire." We must change our methods to better prepare our students mathematically for the technological world.

Use of Groups

Collaborative learning is central to our approach. Each unit in the text includes many group activities. The instructor's edition of the text includes a discussion of active learning. This provides an introduction to group work for the instructor accustomed to a traditional format, and also makes some suggestions that should benefit the collaborative learning veteran.

Prerequisite Skills and Pretest

On page xii is a list of the skills that we consider a prerequisite to this text. Appendix A includes a pretest for prealgebra topics keyed to review exercises. Unit 1 is an introductory unit used to develop collaborative skills and variable concept. You may find this to be an appropriate time to assign review exercises, or you may prefer to assign these exercises just before they are needed in class.

Additional Resources

The *Student Support Manual* (0-201-43457-1) includes worked out solutions for most of the odd problems in the textbook. Solutions to group problems are not included. It also contains appendices for the TI-82 and the TI-86, which follow the same command structure as the TI-83 Appendix in this text.

The Annotated Instructor's Edition (0-201-43451-2) contains teacher notes and answers to all problems in the text. It includes the *Instructor's Resource Manual*, which has exams, active learning hints and suggestions, and the TI-82 and TI-86 appendices.

InterAct Math Tutorial software is available in both Windows (0-201-43455-5) and Macintosh (0-201-43453-9) formats. It provides practice on both prerequisite and other skills encountered in the course.

Information on this and other mathematics reform projects can be found at http://www2.awl.com/mathreform. This site features information on additional resources available for Addison Wesley Longman's reform-oriented titles in developmental mathematics, along with news of national happenings in the mathematics-reform movement, including conferences, workshops, and author events.

Information on graphing calculators and related products is available from Texas Instruments at http://www.ti.com/calc/docs or 1-800-TI-CARES.

Thanks to AWL

We wish to thank Jason Jordan, Publisher; Kari Heen, Senior Editorial Project Manager; Ron Hampton, Managing Editor; Susan Carsten, Designer; Jane Estrella, Production Assistant; and Palmer MacKay, Editorial Assistant at Addison Wesley Longman and Michele Barry, our developmental editor, for their support and encouragement with the project. Their willingness to help us take a rough draft of our concept of the text and to develop it into this edition made the project possible.

Reviewers

Our thanks go to the faculty members who reviewed the drafts of the materials and made suggestions for improvement.

Dianne Adams, *Hazard Community College*

Kathleen Bavelas, *Manchester Community-Technical College*

Frank Cerreto, *The Richard Stockton College of New Jersey*

Virginia Crisonino, *Union County College*

Cindy Daytona, *Fullerton College*

Nancy Henry, *Indiana University Kokomo*

Diane Hillyer, *Manchester Community-Technical College*

Gail Johnson, *Siena Heights College*

Maureen Kelley, *Northern Essex Community College*

Bob Malena, *Community College of Allegheny County*

Joanne Manville, *Bunker Hill Community College*

Allan Robert Marshall, *Cuesta College*

Jan Nettler, *Holyoke Community College*

Ted Panitz, *Cape Cod Community College*

Greg Rosik, *Ridgewater College*

Ellen Schneider, *New Mexico State University*

June Strohm, *Pennsylvania State University–DuBois*

Judy Stubblefield, *Garden City Community College*

Lana Taylor, *Siena Heights College*

Carol Walker, *Hinds Community College*

Class Testers

Additional thanks go to those instructors and students at the listed colleges, who class tested the early drafts and the class test edition. Their comments, corrections, and suggestions led to many improvements in the text.

Mary Beard, *Lexington Community College*

Jim Chesla, *Grand Rapids Community College*

Diane Hillyer, *Manchester Community-Technical College*

Missie Jones, *Owensboro Community College*

Teresa Leyba, *South Mountain Community College*

Barbara Lott, *Maysville Community College*

John Starmack, *Community College of Allegheny County*

Using Formulas and Variables

Upon successful completion of this unit you should be able to:

1. Evaluate formulas using substitution;

2. Complete input-output tables for given formulas;

3. Complete input-output tables for relationships presented as graphs;

4. Recognize and complete patterns;

5. Complete an input-output table for a given formula and plot the values as points on a rectangular coordinate system; and

6. Solve applications.

Variables and Formulas

You have used formulas in previous classes. Most of the letters in these formulas are called **variables** because the values can change or vary.

Your house may be larger or smaller than your neighbor's house. Knowing the measurements from your house, you can use variables and formulas to calculate how much paint you need to purchase, the level of radon exposure, or anything else you need to know. Let's start our discussion of variables by looking at a common formula from geometry.

Area of Rectangles

Examine the following drawing showing the tile floor of a rectangular bathroom in the remodeled home of one of the authors.

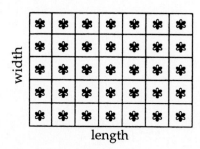

Each tile in the bathroom floor is one square foot in area. Each is one foot long and one foot wide.

1.   How long is the bathroom floor?

2.   How wide is the floor?

3.   How many tiles are needed to cover the floor?

Notice that the total number of tiles is $(7)(5) = 35$. The bathroom floor contains 35 one-foot square tiles and has an area of 35 square feet. Let the variable L represent the length, let W represent the width, and let A represent the area of the rectanglar floor. We write the formula for the area as $A = L \cdot W$.

 The pencil-and-paper icon marks a problem or activity that requires your written response.

Let's calculate the area by substituting the value of 7 feet for the length and 5 feet for the width.

$$A = L \cdot W = (7 \text{ feet})(5 \text{ feet}) = 35 \text{ square feet}$$

Here is another rectangular bathroom floor constructed from one-foot square tiles.

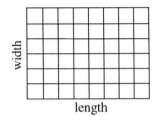

width

length

1.   How long is the floor?

2.   How wide is the floor?

3.   How many total tiles are in the floor?

4.   Calculate the area by substituting the values for the length and width in the formula: $A = L \cdot W$.

 You priced labor because you decided not to lay the tile yourself. If the labor is $3.50 per square foot, what will be the total bill for labor?

The cost of $3.50 per square foot can also be written as $\dfrac{\$3.50}{\text{square foot}}$. The total area to be tiled is 48 square feet. Multiply the area by the cost per square foot.

$$48 \text{ square feet} \cdot \frac{\$3.50}{\text{square foot}} = \$168$$

By using units throughout the problem, you know to multiply instead of divide by the cost per square foot. When you use the appropriate operation, the remaining units are reasonable and correct.

 The magnifying-glass icon indicates examples that you should examine carefully.

You Try It

1. Find the number of square yards of carpet needed to cover the floor of a room that is 5 yards long and 4 yards wide. Sketch a picture, and then calculate the area using the formula $A = L \cdot W$.

2. Find the area of a backyard that is 35 yards long and 20 yards wide. How much grass seed will you need to seed the yard if one pound of seed covers 25 square yards?

Order of Operations

In your group, each member should evaluate $3 + 5 \cdot 4$ writing the answer on paper before sharing with other group members.

Does everyone have the same answer? Or do some group members have the answer 32 while others have 23? Who is correct?

Imagine the confusion that would occur if a simple problem such as $3 + 5 \cdot 4$ had two different answers. To avoid chaos, the order of operations agreement was developed. According to the agreement, $3 + 5 \cdot 4 = 23$. Let's review the order of operations agreement to determine why 23 is the correct answer.

Order of Operations Agreement

1. Perform operations inside grouping symbols: parentheses ( ), brackets [ ], absolute value | |, and in the numerator or denominator of fractions.
2. Simplify expressions with exponents.
3. Multiply and divide left to right as they occur.
4. Add and subtract left to right as they occur.

There are many memory devices available to help you remember the order of operations agreement. One memory device is to use the initials of the operations including arrows as a reminder to proceed left to right when multiplying and dividing or adding and subtracting: $PE \underset{\longrightarrow}{MD} \underset{\longrightarrow}{AS}$. For the expression above, multiply before you add.

 The lizard icon marks a problem or activity designed to be worked in groups of 3–5 students with a report due according to your instructor's directions.

1.   Simplify: $(12-7)^2(5) - 12 \div 6$

$(12-7)^2(5) - 12 \div 6$
$(5)^2(5) - 12 \div 6$ *Perform operations inside parentheses*
$25(5) - 12 \div 6$ *Simplify expressions with exponents*
$125 - 2$ *Multiply and divide, left to right*
123 *Subtract*

2.   Simplify: $6 \div 2[4-(7-6)] - 1 + 4$

$6 \div 2[4-(7-6)] - 1 + 4$
$6 \div 2[4-(1)] - 1 + 4$ *Perform operations inside parentheses*
$6 \div 2[3] - 1 + 4$ *Perform operations inside brackets*
$3[3] - 1 + 4$ *Divide since it appears first, left to right*
$9 - 1 + 4$ *Multiply*
$8 + 4$ *Subtract*
12 *Add*

You Try It

Use the order of operations agreement to simplify each of the following expressions. Identify the operation you are performing in each step.

1. $\dfrac{12}{4} - 3$

2. $5 + 3 \cdot 4$

3. $4 + 3^2$

4. $(4+3)^2$

5. $(5+3) \cdot 4^2$

6. $\dfrac{7-4}{3^2}$

Use the order of operations with pencil and paper and your graphing calculator to simplify these expressions. If the answers do not agree, review your work and calculator screen to correct any errors.

7. $\left(\dfrac{1}{2}\right)^2 - \left(\dfrac{3}{4} - \dfrac{1}{2}\right)$

$(1 \boxed{\div} 2)\boxed{x^2}\boxed{-} (3 \boxed{\div} 4 \boxed{-} 1 \boxed{\div} 2)$

8. $\dfrac{10+2}{6^2 - 30} \div (62 - 60)$

$(10+2) \boxed{\div} (6 \boxed{x^2} \boxed{-} 30) \boxed{\div} (62 \boxed{-} 60)$

9. $35 - 2(7-3)^2 + 6$

$35 \boxed{-} 2 \boxed{\times} (7 \boxed{-} 3) \boxed{x^2} \boxed{+} 6$

10. $3^2 \cdot 2 - 3 + 4$

$3 \boxed{x^2} \boxed{\times} 2 \boxed{-} 3 \boxed{+} 4$

Perimeters of Rectangles

Let's look at the first bathroom floor again. We now are interested in how many feet of baseboard trim we would need along the walls to cover the edges of the tile floor.

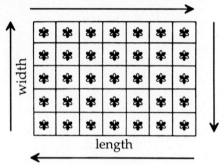

The length of the room is 7 feet and the width is 5 feet. To cover the edges of the room in trim, we would have to cover the base of the wall for all the edges. Imagine walking around the room, starting at the top left hand corner. Following the arrows, this is two lengths and two widths.

Trim needed =	7 feet + 5 feet + 7 feet + 5 feet
=	7 feet + 7 feet + 5 feet + 5 feet
=	2 (7 feet) + 2 (5 feet) = 14 feet + 10 feet
=	24 feet

The distance around an object is called perimeter.

Let's write the formula for the perimeter of the rectangle following the pattern above. Use P for perimeter, L for length, and W for width.

$$P = L + W + L + W$$

$$= L + L + W + W$$

or

$$P = 2L + 2W$$

Use the formula to calculate the amount of baseboard trim needed.

$P = 2L + 2W$
$P = 2(7 \text{ ft}) + 2 (5 \text{ ft})$ *Use the order of operations agreement*
$P = 14 \text{ ft} + 10 \text{ ft} = 24 \text{ ft}$

So far, we've examined two formulas concerning rectangles. Both are written in terms of length and width. Complete the two formulas below.

Perimeter of a Rectangle: $P =$ _____

Area of a Rectangle: $A =$ _____

 How are they the same, and how are they different?

Both formulas have three letters, called variables. The perimeter formula also contains the number two. When a number is used in a formula, it is called a **constant**. Does the formula for the area of a rectangle contain any constants? Are there any other differences between the two formulas?

You should remember that perimeter is a length and is measured in linear units. Area covers a surface and is measured in square units, as in the tile floor problem. Note that we add units when calculating perimeter but multiply units when calculating area.

You Try It

1. Calculate the perimeter of a bathroom floor that is 6 feet wide by 8 feet long.

2. What is the number of feet of trim needed along the walls of a room that is 15 feet long and 12 feet wide?

3. Find the number of linear feet of chain link fence needed to enclose a yard that is 35 yards long and 20 yards wide. Remember to change the yards to feet before you find the perimeter.

Evaluation of Formulas Using Substitution

In the previous section, we found that the formula for the perimeter of a rectangle was $P = 2L + 2W$. We used the variables P, L, and W to represent the perimeter, length and width.

 What is the perimeter of each of the following rectangles with length by width?
1.   5 in by 4 in
2. $5\frac{1}{3}$ yards by $3\frac{1}{4}$ yards
3.   34.75 m by 15.45 m

First write the formula for perimeter, $P = 2L + 2W$.

1.   To find the first perimeter, substitute 5 inches for L and 4 inches for W:

$$P = 2L + 2W = 2(5 \text{ in}) + 2(4 \text{ in}) = 10 \text{ in} + 8 \text{ in} = 18 \text{ in}$$

2.   Substitute $5\frac{1}{3}$ yards for L and $3\frac{1}{4}$ yards for W:

$$P = 2L + 2W = 2\left(5\frac{1}{3} \text{ yards}\right) + 2\left(3\frac{1}{4} \text{ yards}\right)$$

$$= 2\left(\frac{16}{3} \text{ yards}\right) + 2\left(\frac{13}{4} \text{ yards}\right)$$

$$= \frac{32}{3} \text{ yards} + \frac{13}{2} \text{ yards}$$

$$= \frac{64}{6} \text{ yards} + \frac{39}{6} \text{ yards} = \frac{103}{6} \text{ yard} = 17\frac{1}{6} \text{ yards}$$

3.   Substitute 34.75 m for L and by 15.45 m for W:

$P = 2L + 2W$
$P = 2(34.75 \text{ m}) + 2(15.45 \text{ m})$
$P = 69.50 \text{ m} + 30.90 \text{ m}$
$P = 100.4 \text{ m}$

You Try It

1. Find the perimeter of a rectangle that is 17 cm by 8 cm.

2. Find the perimeter of a rectangle with measurements of $4\frac{4}{5}$ ft by $3\frac{2}{3}$ ft.

3. Find the perimeter of a rectangle that is 21.03 yd long and 7.19 yd wide.

Area of Rectangles Revisited

Earlier, we found that the formula for the area of a rectangle was $A = L \cdot W$. We used the variables A, L, and W to represent the area, length and width.

Calculate the area of rectangles with length by width as follows:
1. 5 in by 4 in
2. $5\frac{1}{3}$ yards by $3\frac{1}{4}$ yards
3. 34.75 m by 15.45 m

First, write the formula for area, $A = L \cdot W$.

1. To find the first area, substitute 5 inches for L and 4 inches for W:

   $A = L \cdot W = (5 \text{ in})(4 \text{ in}) = 20 \text{ in}^2 = 20 \text{ square inches}$

2. Substitute $5\frac{1}{3}$ yards for L and $3\frac{1}{4}$ yards for W:

   $$A = L \cdot W = \left(5\frac{1}{3}\text{ yards}\right)\left(3\frac{1}{4}\text{ yards}\right) = \left(\frac{16}{3}\text{yards}\right)\left(\frac{13}{4}\text{yards}\right)$$

   $$= \frac{52}{3}\text{ yards}^2 = 17\frac{1}{3}\text{ yards}^2 = 17\frac{1}{3}\text{ square yards}$$

3. Substitute 34.75 m for L and 15.45 m for W:

   $A = L \cdot W = (34.75 \text{ m})(15.45 \text{ m}) \approx 536.89 \text{ m}^2 = 536.89 \text{ sq. meters}$
   The $\approx$ means approximately and is used when the result is rounded.

You Try It

1. Calculate the area of a rectangle with measurements of 17 yards by 8 yards.

2. Determine the area of a rectangle that measures $4\frac{4}{5}$ ft by $3\frac{2}{3}$ ft.

3. Calculate the area of a rectangle that is 21.03 yd long and 7.19 yd wide.

Sales Tax

The sales tax in Mason County is six and three-fourths percent $\left(6\frac{3}{4}\%\right)$ of the purchase price.

Calculate the sales tax on purchases of
1. $12.78   2. $101   3. $333.33

After you convert the percent to a decimal, the formula for sales tax with variables S and P is $S = 0.0675\,P$.

1.   To find the first sales tax, substitute $12.78 for P:

$S = 0.0675\,P = 0.0675\,(\$12.78) \approx \$0.86$

2.   To find the second sales tax, substitute $101 for P:

$S = 0.0675\,P = 0.0675\,(\$101) \approx \$6.82$

3.   To find the last sales tax, substitute $333.33 for P:

$S = 0.0675\,P = 0.0675\,(\$333.33) \approx \$22.50$

You Try It

1. Use the Mason County sales tax to calculate the sales tax on purchases of $2.28 and $399.

2. Use your local sales tax rate to determine the sales tax on total sales of $6.25 and $564.

Degrees Celsius to Degrees Fahrenheit

Two scales are used for temperature in the USA—Celsius and Fahrenheit. Most people in the United States are more familiar with temperatures reported in Fahrenheit, so they convert from Celsius to Fahrenheit. To convert from degrees Celsius to degrees Fahrenheit, you multiply the Celsius temperature by nine-fifths then add 32.

1. Write the formula for changing from degrees Celsius to degrees Fahrenheit using variables.

2. Convert the following degrees Celsius to degrees Fahrenheit:

   a) $0°C =$

   b) $42.5°C =$

   c) $74\frac{1}{3}°C =$

   d) $100°C =$

   e) $124.8°C =$

3. Which of the above temperatures represents the temperature at which water boils?

4. Which represents the temperature at which water freezes?

Group Tip: Show all the group work on your own paper so you will have results to review.

Simple Interest

Interest = (Principal)(Rate)(Time) shortened to $I = PRT$ is another formula that you may have encountered in your studies. This formula is used to calculate the simple interest earned by an investment. Today, most banks use compound interest rather than simple interest. Later in this text, you will work problems with compound interest.

Simple interest, I, is found by multiplying the value of the principal, P, by the rate of interest, R, and the time invested, T. The interest and principal are usually expressed in dollars, the rate as a percent per year, and the time in years.

1. Calculate the simple interest on an investment of $2500 at 3% for 3 months.

$P = \$2500$

$R = 3\%$

$T = 3 \text{ months} = 3 \text{ months} \left(\dfrac{1 \text{ year}}{12 \text{ months}} \right) = \dfrac{1}{4} \text{ year}$

To use the formula, we must change the 3 months into the equivalent part of a year. We can do this by multiplying 3 months by the unit fraction in which the numerator is 1 year and the denominator is the equivalent number of months.

$$I = PRT = (\$2500)\left(\frac{3\%}{\text{year}} \right)\left(\frac{1}{4} \text{ year} \right) = (\$2500)(0.03)\left(\frac{1}{4} \right) = \$18.75$$

2. What will you earn if you invest $2500 at 3% simple interest per year for 3 years?

The principal, P, is $2500. R is the decimal value of 3% per year, and T is 3 years. You will earn $225 as shown below.

$$I = PRT = (\$2500)\left(\frac{3\%}{\text{year}} \right)(3 \text{ year}) = (\$2500)(0.03)(3) = \$225$$

Remember, if you use units consistently, you will often prevent many careless errors.

You Try It

1. Find the interest earned for an investment of $300 for 2 years at 4%.

2. Find the interest earned for $2,000,000 invested for 10 years at 6.75%.

Input-Output Tables

Sometimes we need to evaluate a formula for several different values of a variable. A **variable** is a symbol (usually a letter) that is used to represent an unknown number or any one of several numbers.

When we evaluate a formula for several values, we can organize the data in a table called an input-output table. The inputs for the table will be selected values for the variable. We calculate the outputs by substituting the inputs into the formula. These tables are often useful in problem solving when what-if questions arise.

Distance vs. Time

To calculate the distance traveled by an object, you multiply the rate of the object by the time traveled. This is written $D = RT$, where D is the distance, R is the rate, and T is the time.

Assume that you are traveling on Interstate 5 between Sacramento and Seattle traveling at a constant rate of 65 miles per hour. Use the formula for distance and substitute the 65 miles per hour for the rate, R, to obtain $D = 65T$.

Suppose you want to calculate the distance you would travel in 1, 2, 3, 4, and 5 hours. You could substitute each number in the formula as before. However, let's look at a more efficient way to do this using an input-output table.

Look at the following table. The two columns in the first row are labeled T for time, which is the input, and D for distance, which is the output. The time values that we choose to use in our calculations are written in the remaining rows.

T	D = 65T
1 hr	
2 hr	
3 hr	
4 hr	
5 hr	

The following table contains the formula for distance and the calculations of the distance traveled for the given times. When completing input-output tables, use units for the first output that you calculate to verify that the units are correct.

T	D = 65 T
1 hr	(65 miles/hr)(1 hr) = 65 miles
2 hr	(65)(2) = 130 miles
3 hr	(65)(3) = 195 miles
4 hr	(65)(4) = 260 miles
5 hr	(65)(5) = 325 miles

You Try It

Label the input and output columns in the following table that has a different rate and different times. Complete the table to calculate the distance traveled for the given times.

T	D = 55 T
1.5 hr	
2.3 hr	
3.7 hr	
4.8 hr	
5.0 hr	

Review of Integer Arithmetic

Integer arithmetic is an important topic in prealgebra classes. The **integers** are
…–4, –3, –2, –1, 0, 1, 2, 3, 4, …

Integers are often shown on the number line as below:

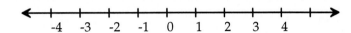

In your group, review the rules that you learned and for each statement below,
give rules and examples to help you remember how to perform each integer
arithmetic operation. Your group must include negative integers when writing
your rules and examples.

1. $a + b$, where a and b are integers.

2. $a - b$, where a and b are integers.

3. $a \cdot b$, where a and b are integers.

4. $a \div b = \dfrac{a}{b}$, where a and b are integers and $b \neq 0$.

5. a^n, where a is an integer and n is a whole number.

6.   Several integers to add and subtract.

Before completing this *You Try It*, confirm your rules as directed by your
instructor.

You Try It

Perform the indicated operations.

1. $(-5) + (-4)$

2. $(-5) - (-4)$

3. $(-3)(5)$

4. $\dfrac{-12}{-3}$

5. $(-2) - (-3) + (-7)$

6. $(-2)^3 (-3)^2$

7. $-2^2 - 5 + 3$

8. $(-3)^2 \div 3(2)$

9. $\dfrac{9}{5}(-30) + 32$

10. $-5 + (+3)$

11. $(-5) + (+7)$

12. $(-5) + (-7) - (+3) - (-8)$

Perimeter of Rectangles Revisited

Earlier, we used the formula $P = 2L + 2W$ to find the perimeters of rectangles after measuring lengths and widths as follows: 5 in by 4 in; $5\frac{1}{3}$ yards by $3\frac{1}{4}$ yards; 34.75 m by 15.45 m. We could also have done this by completing the following input-output table.

L	*W*	*P* = 2*L* + 2*W*
5 in	4 in	
$5\frac{1}{3}$ yds	$3\frac{1}{4}$ yds	
34.75 m	15.45 m	

Complete the table and compare your answers with the ones from the previous section.

Celsius and Fahrenheit Temperature Revisited

We've discussed the two scales used for temperature in the USA—Celsius and Fahrenheit. We used the formula $F = \frac{9}{5}C + 32$ to convert from degrees Celsius to degrees Fahrenheit. Use an input-output table to convert from degrees Celsius to degrees Fahrenheit.

C	$F = \frac{9}{5}C + 32$
$-80°$	
$-45°$	
$0°$	
$37°$	
$42.5°$	
$74\frac{1}{3}°$	
$100°$	

Simple Interest Revisited

Complete the following table to compute the simple interest for the given investments. Remember that the time must be converted to years when you compute simple interest.

P	R	T	$I = PRT$
$500	6%	2 months	
$3500	2.99%	6 months	
$5000	4 ½%	90 days	
$2,000,000	7.99%	3 years	
$20,000,000	4%	30 days	

You will use input-output tables throughout the textbook. Later in this unit, you will generate ordered pairs on input-output tables and plot them on rectangular coordinate systems.

Graphs Instead of Formulas

Sometimes the relationship between variables is given as a graph instead of a formula. In these cases, we must read the graph to find the output for a particular input. Popular magazines and newspapers use graphs to communicate information quickly and concisely.

> A graph has input values on the horizontal axis and output values on the vertical axis.

The following graph demonstrates the total federal debt in billions of dollars as a function of the fiscal year.

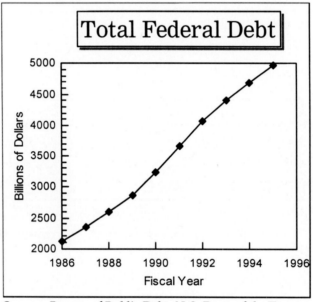

Source: *Bureau of Public Debt, U.S. Dept. of the Treasury*

When you look at this graph you may notice that the fiscal year is the input and the debt in billions of dollars is the output. For an input of fiscal year 1992, how could you find the debt in billions of dollars?

Put your finger on the 1992 and move it up to the graph. When your finger touches the graph, move it to the left until you locate the value (approximately 4100). For fiscal year 1992, the federal debt was approximately 4100 billion dollars or four trillion, one hundred billion dollars.

Now suppose you want to know when the federal debt was 2600 billion dollars. Put your finger on 2600 and move it to the right until you are on the graph. Move your finger down until you locate the value which is about 1988. Therefore, as we interpret the graph, for fiscal year 1988, the federal debt was 2600 billion dollars.

You Try It

1. Find the federal debt for each of the following fiscal years:

   a)   1994

   b)   1989

   c)   1993

   d)   1986

2. By approximately how many billion dollars did the federal debt increase between 1987 and 1988? Between 1990 and 1991? For which pair of years was the increase larger?

3. Find the year in which the federal debt was closest to

   a)   2350 billion dollars

   b)   4692 billion dollars

Blood Sugar

The following graph shows the level of glucose in the blood of a normal patient when tested before and following a meal high in carbohydrates. The vertical line at 0 represents the time when the patient ate.

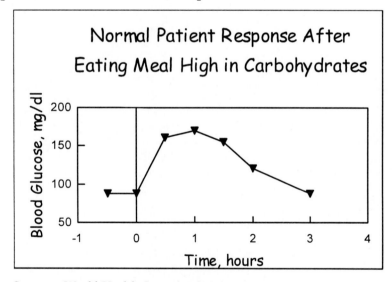

Source: *World Health Organization*

We can say that the person's blood glucose depends on time or that blood glucose is a function of time. We can use an input-output table to show the relationship between blood glucose and time. Complete the following table.

Look at the rows with outputs. Do you agree with the numbers? Your interpretation of a graph may be slightly different from someone else's. Compare your answers with those of your group and try to reach a consensus.

Time, hours	Blood Glucose
−0.5	85 mg/dl
0	
0.5	
1	
1.5	
	130 mg/dl
3	85 mg/dl

You Try It

The following graph shows the life expectancy as a function of current age.

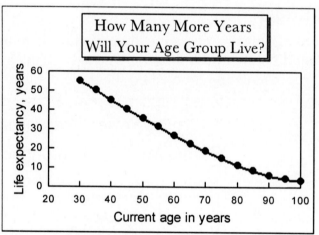

Source: *U.S. Census Bureau.*

1. Complete the input-output table below.

Current Age	Life Expectancy
35	
40	
50	
65	
95	
	10
	25
	50

2. Compare the total life expectancy of a 30-year-old person with the total life expectancy of a 60-year-old person.

Patterns

Your life is filled with patterns. Some patterns are obvious while others are very subtle. Observing patterns will help you solve many problems during this course and in later mathematics courses. To begin our observation, let's look at some patterns you may already recognize.

Complete the following patterns:

1. 2, 4, 6, 8, _____ , _____ , _____

2. 1, 4, 9, 16, _____ , _____ , _____

3. Feb., Mar., Apr., _____ , _____ , _____

4. A, B, C, _____ , _____ , _____

As you observed, to complete the patterns above you needed to recognize the even numbers, perfect squares, the months of the year, and the alphabet. Now try completing the following pattern:

5. 10, 14, 18, 22, _____ , _____ , _____

If you realized that you needed to add 4 to each term to obtain the next one, you were on the right path. After an initial look for obvious patterns, we begin our search for a pattern by looking for constant differences.

Second Number – First Number = 14 – 10 = 4
Third Number – Second Number = 18 – 14 = 4
Fourth Number – Third Number = 22 – 18 = 4

We often show our work like: 10, 14, 18, 22, _____ , _____ , _____

4   4   4

Since all the differences are the same, the subtle pattern becomes obvious. We must add the constant 4 each time to obtain the following number.

You Try It

Complete the following patterns; remember to look for obvious patterns first.

1. 5, 11, 17, ____ , ____ , ____

2. 200, 220, 240, ____ , ____ , ____

3. 17, 22, 27, ____ , ____ , ____

Patterns in Input-Output Tables

Often patterns will help you complete input-output tables. For input-output tables (I-O tables), you must concern yourself with observable patterns for both the input and output. You must first confirm that there is a constant difference between each x-value. For example, the first step for this problem is to find the differences between each input, x.

$0 - (-1) = 1$
$1 - 0 = 1$
$2 - 1 = 1$
$3 - 2 = 1$
$4 - 3 = 1$

x	y
−1	−8
0	−11
1	−14
2	
3	
4	

There is a constant difference of 1 between each x-value. You can now observe the output, y, for a pattern to complete the input-output table as we did earlier. While we are examining patterns, always remember that the output y depends on the input x.

Here is another input-output table to complete, if possible.

x	−9	−12	−15	−18	−21	−24
y	31	32	33			

Your completed work might look like the work below:

	-3	-3	-3	-3	-3
x	-9	-12	-15	-18	-21	-24
y	31	32	33	34	35	36
	1	1·	1	1	1

Be careful, though. If you do not have constant differences in the input values, you cannot blindly look for a pattern in the output values.

1.   Examine the following table and determine the differences between each pair of input values and each pair of output values.

x	*y*
-7	-30
-2	-5
3	20
8	45
18	95
23	120

2.   Is there a pattern in the outputs? Why or why not? Later you will learn other ways to determine the output value to complete the table.

You Try It

If possible, complete the following I-O tables by observing patterns.

1.

x	*y*
-3	-21
-1	-11
1	-1
3	
5	

2.

x	*y*
0	3
2	-1
4	-5
6	
8	

3.

x	*y*
-1	5
0	2
1	-1
2	
5	

To Burn or Not To Burn

Stephanie and her friends are debating the use of suntan lotion to prevent burning. They learned that an SPF of 2 blocked 50% of the UV rays, an SPF of 4 blocked 75% of the UV rays, while an SPF of 6 blocked $83\frac{1}{3}$% of the UV rays. If Stephanie was told she needed to block at least 90% of the UV rays, what SPF should she use?

1.   Which information should you use as input? Why?

2.   Which should you use as output? Why?

3.   Complete the following input-output table using the numbers above.

Differences			
SPF			
% of UV rays blocked			
Differences			

4.   Is there a constant difference between the inputs?

5.   Is there a constant difference between the outputs?

6.   After changing all the values for % of UV rays blocked to simplified fractions, complete the following table:

SPF			
Fractional amount of UV rays blocked			

7.   Compare the SPF value to the denominator of the corresponding fraction.

8.   Compare each numerator with the denominator of the same fraction, is there a pattern?

9.   What fraction would correspond to an SPF of 8? Of 15? Of 30?

10.   What % of the UV rays blocked corresponds to an SPF of 8? Of 15? Of 30?

11.   What SPF would you recommend Stephanie use?

 Group Tip: Wisdom is knowing not only when to speak your mind, but also when to mind your speech. Groups are more often successful when everyone participates actively and when each member listens carefully to other students' ideas.

Input-Output Tables to Plot Points

One application of input-output tables is to generate sets of ordered pairs to plot on a rectangular coordinate system. Consider the graph paper below.

Notice that the horizontal line is labeled as the *x*-axis and the vertical line as the *y*-axis. The origin has coordinates (0, 0) and is located at the intersection of the *x*-axis and *y*-axis.

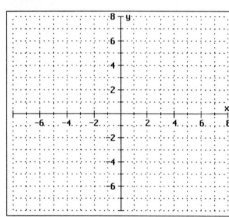

Each point has both an *x*-coordinate and a *y*-coordinate. Positive *x*-coordinates are plotted to the right of the *y*-axis while negative *x*-coordinates are to the left of the *y*-axis. Positive *y*-coordinates are plotted above the *x*-axis while negative *y*-coordinates are below the *x*-axis.

Let's plot the points defined by this table on the graph paper above.

To plot the first point of the table, (0, 0), draw the point at the origin.

x	y
0	0
3	2
2	–3
–3	2
–2	–3

Beginning at the origin, the point (3, 2) would be three units to the right and two units up. From the origin, the point (2, –3) would be two units to the right and three units down. The point (–3, 2) would be three units to the left and two units up from the origin, while (–2, –3) would be two units to the left and three units down from the origin.

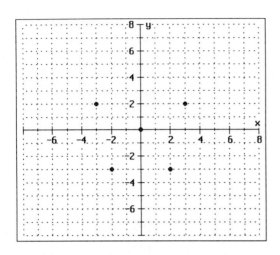

The points are plotted at the left. Label each point using the notation (input value, output value). Compare with your graph.

Here is an input-output table for the formula $y = 3x - 4$.

The points from the values in the input-output table are plotted below.

x	$y = 3x - 4$
–1	–7
0	–4
1	–1
2	2
3	5

Note that positive x-coordinates are to the right of the y-axis while negative x-coordinates are to its left. Positive y-coordinates are above the x-axis while negative y-coordinates are below.

Label each point.

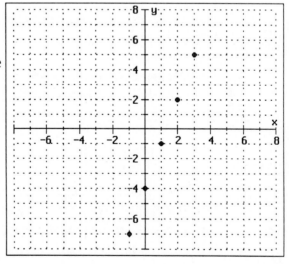

Plotting Points on the TI-83

We can use the TI-83 to plot the same points.

x	−1	0	1	2	3
$y = 3x - 4$	−7	−4	−1	2	5

This is the same table written horizontally rather than vertically.

To plot the points on the TI-83 graphing calculator, press STAT to display the screen shown here.

Press 1 or ENTER to edit the entries in the **LIST**.

Your calculator screen should look like this one.

If there are entries in the **LIST**, press ▲ until the **L1** is highlighted. Then CLEAR ENTER to erase all entries in the first list. Press ▶ ▲ CLEAR ENTER to erase **L2**. Press ◀ until **L1(1)=** shows at the bottom of the screen.

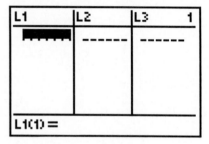

Once there are no entries in **L1** and **L2**, type the first *x*-value, −1. Press ENTER. Continue this process until you have entered the last *x*-value.

Press the right arrow, ▶. This will move the cursor to the top of the next list, **L2**. Now enter the *y*-values in the same order. The TI-83 screen should look similar to this one.

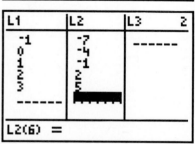

There must be the same number of entries in **L1** as there are in **L2** when you plot data as points; if not, an error message will appear in the window.

Now press Y=. The screen should look like this one.

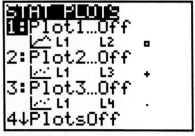

If not, press CLEAR to erase the first equation. Press ▼, then CLEAR until all equations are erased.

Press 2nd Y= to display **STAT PLOTS**. Plots 1, 2, and 3 should all display **Off** at this time. If any plots display **On**, press 4 ENTER. This will return you to the **HOME** screen so you will need to press 2nd Y= to display **STAT PLOTS** again.

Press 1. Although the **Off** is highlighted, the cursor should be flashing at the **On**. If so, press ENTER to turn **Plot 1** on. The **Type:** should be ⌐∷ (Scatterplot). For this plot, the **Xlist** indicates where the *x*-values are located (**L1**). The **Ylist** indicates where the *y*-values are located (**L2**). Choose any of the three marks, or use the one your instructor suggests.

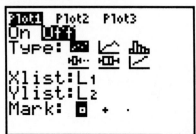

If the type is not ⌐∷ , arrow down and press ENTER to select it. Once your TI-83 screen matches the screen on the right, press WINDOW to define the portion of the coordinate system that is displayed on the screen.

To plot the given ordered pairs on graph paper you would consider the largest and smallest values for the axes. (You will practice this skill in the next group problem.) The same is true for the graphing calculator.

1. What is the smallest *x*-value in the preceding table?

2. What is the largest *x*-value?

3. What is the smallest *y*-value?

4. What is the largest?

To see all the points, it is customary to choose the maximum values on the axes slightly larger than the largest values in the table. Similarly, the minimum values we choose to be slightly smaller than the table values. Let's plot the points on the portion of the coordinate system between –2 and 4 on the horizontal axis and –10 and 10 on the vertical axis. We will use –2 for the **Xmin** value and 4 for the **Xmax** value in the viewing window. We will use –10 for the **Ymin** and 10 for the **Ymax** values.

Although the scale is not critical at this time, we may wish to change the scale to 1 for both the **Xscl** and the **Yscl**.

Press WINDOW. Press (-)2 ENTER to change the **Xmin**. Press 4 ENTER for **Xmax**. Continue this process until the screen matches the one displayed here. Be careful to press the negative, (-), not the minus, –, when entering these values.

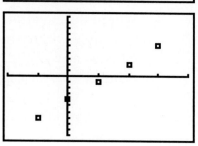

Now press GRAPH to view the plot.

Often during this course and later ones, you will plot points to determine the shape of a graph.

 What shape does the graph of your data and the graph above suggest? Look at the points on the graph. Would the points form a line if they were connected?

Each of the following graphs shows the points connected with a line.

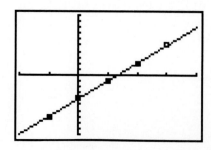

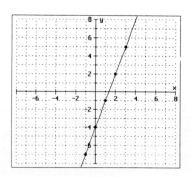

Complete the following table and plot the points on the graph provided.

x	$y = x^2 - 3$
–3	
–2	
–1	
0	
1	
2	
3	

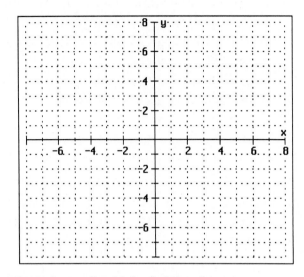

If you plotted the points on your TI-83, they would look like this:

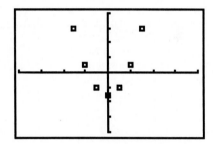

If these points are connected, will they form a line? No, this graph is called a parabola, and is shown here.

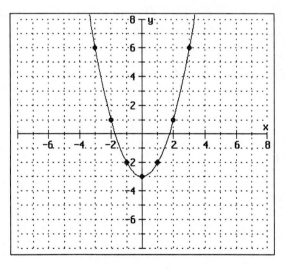

Complete the following table and plot the points.

x	$y = \sqrt{(x+4)}$
-4	
-3	
0	
2	
5	
7	

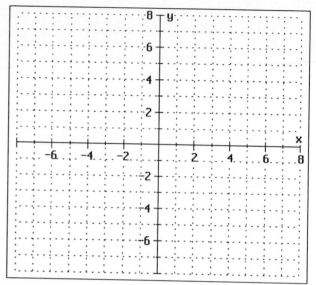

Now plot the points on your TI-83. Here is the way the screen should look.

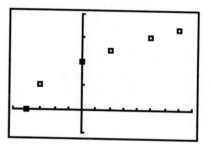

The points from this square root table would form the following graph:

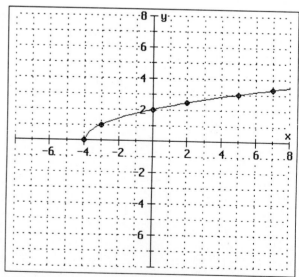

In this section, we have seen how input-output tables can be used to generate ordered pairs for graphs. We will look at a variety of such graphs in the problem set.

Big and Little Numbers

Plot the values from the following input-output tables. Carefully label the axes with x, y, and indicate the scale.

1.

x	y
10	0.5
20	1
−10	−0.5
−20	−1
25	1.25

2.

x	y
2	−32
4	−32
5	−32
6	−32
7	−32

3.

x	y
−18	20
−6	−12
0	24
12	−6
6	12

4.

x	y
−5	15
−10	−10
25	30
15	−30
0	5

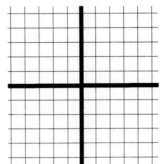

5.   Complete the values that you would need to plot the points (–5, 25) and (30, –5) on your TI-83.

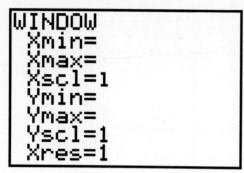

```
WINDOW
 Xmin=
 Xmax=
 Xscl=1
 Ymin=
 Ymax=
 Yscl=1
 Xres=1
```

6.   **a)**   Plot the points (1, –3) and (–3, 2).

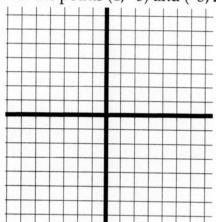

b)   With a straight edge or ruler, sketch a line that passes through the two points.

c)   Does the point (5, –8) lie on the line?

d)   Estimate what *y*-value corresponds to an *x*-value of –2.

7.   If your group finishes early, plot the values from the input-output tables in **1–4** on your calculator.

Project

Apple Carpet Company

TO: Team Members

FROM: Your Supervisor

Our group, Apple Carpet Company, has been asked to submit a bid for Xeineth Community College. They wish to install new carpet in three rooms of the Mathematics-Science Building. The Superintendent of Building and Grounds, your instructor, will direct you to the rooms.

Your first task is to estimate the length and width of each room.

You are to find the length and width of each of the rectangular rooms. Once you have these measurements, you will compute the area of each in square yards and use the following table to give three estimates for each room.

Carpet Grade	Price Per Sq. Yard	Installation Cost Per Sq. Yard
Heavy Traffic	$12.95	$2.50
Commercial	$ 9.84	$2.50
Industrial	$7.63	$2.50

Your report should be in the form of a bid to Dr. Taylor, Dean of Business Affairs, Xeineth Community College.

Summary

During your study of this unit, you have:

1. Evaluated formulas by substituting a number for the variable. While this works for any formula, you have learned the following formulas:

    Area of a rectangle: $A \cdot L = W$
    Perimeter of a rectangle: $P = 2L + 2W$
    Interest = (Principal)(Rate)(Time): $I = PRT$
    Fahrenheit $= \dfrac{9}{5} \cdot$ Celsius $+ 32$: $F = \dfrac{9}{5} \cdot C + 32$
    Distance = (Rate)(Time): $D = RT$

2. Completed input-output tables for given formulas;

3. Reviewed arithmetic of whole numbers, fractions, decimals, and integers;

4. Completed input-output tables from published graphs;

5. Recognized and completed patterns in input-output tables;

6. Completed an input-output table for a given formula and plotted the values as points on a coordinate system;

7. Plotted points on graph paper and on the TI-83;

8. Solved application problems involving input-output tables and evaluation of formulas; and

9. Used the following TI-83 features:

    [STAT] to enter data and **STATPLOT** to turn on a plot
    [WINDOW] to set up the viewing window
    [GRAPH] to view the graph

Unit 1 Problems for Practice

A. Complete the following problems.

1. **a)** $523 + 496$ **b)** $8235 - 421 \cdot 3$
   c) $-173 \div 3$ **d)** $5281 \div 153$
   e) $-396 + 27 \cdot 7$ **f)** $721 \cdot 31 - 883 \cdot 22$

2. **a)** $56 \cdot 18 + 53 \cdot 8 - 3357$ **b)** $1358 \div 9 + 82 \cdot 15$
   c) $-780 \div 26 + 113$ **d)** $-57 \cdot 17 + 540 \div 9$

3. Simplify:
   a) $17 - 5(11 - 23) + 61$
   b) $15 - 12(-11) - (31 - 50)$
   c) $4 + 2(6 - 15) - 5(-11)$
   d) $1256 - 16(3001 - 1220)$
   e) $13(-2 + 17) - 11(21 - 15 \cdot 7)$

4. Evaluate:
   a) $5.24 + 3.77 - 5.16$ **b)** $1.73 \cdot 3.5$
   c) $5.87 - 2.9551$ **d)** $1.81(-2.7)$
   e) $-2.81 \div (-0.3)$

5. Evaluate:
   a) $1.2 + 5.09 - 2.2$ **b)** $-1.24(-2.3) + 0.21$
   c) $5 - (8.2 - 1.4) + 2 \cdot 1.7$ **d)** $23.5 \cdot 3.21$
   e) $-5.32(-1.5) + 1.2(-0.3)$

6. Perform the indicated operation:
   a) $\dfrac{1}{3} + \dfrac{2}{5}$ **b)** $2 + \dfrac{2}{3} + \dfrac{1}{10}$

   c) $\dfrac{5}{8} - \dfrac{1}{4}$ **d)** $\dfrac{7}{16} + 2\dfrac{1}{8}$

7. Perform the indicated operation:
   a) $\dfrac{1}{2} + 3\dfrac{5}{8}$ **b)** $\dfrac{2}{3} \div \dfrac{3}{10}$

   c) $\dfrac{87\frac{1}{2}}{100}$ **d)** $2\dfrac{7}{8} \cdot 3\dfrac{1}{16}$

8. Perform the indicated operation:

 a) $\dfrac{31\frac{1}{4}}{100}$

 b) $2\dfrac{2}{3} \div \dfrac{8}{5}$

 c) $\dfrac{1}{8} - \dfrac{3}{4}$

 d) $\dfrac{83\frac{1}{3}}{100}$

9. Change the following percentages to decimals:
 a) 12.57%
 b) 2.987%
 c) 324.7%
 d) 230%
 e) 78.23%
 f) 2.01%

10. Change the following decimals to percentages:
 a) 0.78911
 b) 0.10005
 c) 0.4892
 d) 0.001205
 e) 2.576
 f) 33

11. Evaluate the following:
 a) 20% of 35
 b) 12% of $250
 c) 2% of 980
 d) $7 is what percent of $225?
 e) $18 is what percent of $200?
 f) 12.5% of $2.95 is what?

B. Complete the following problems.

12. Calculate the perimeter of:

 a)

 b)

13. Calculate the area of the figures in **12**.

14. Calculate the perimeter of:

a)

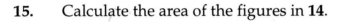

b)

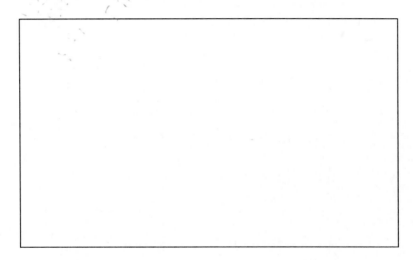

15. Calculate the area of the figures in **14**.

16. Measure and label the sides of this rectangle to the nearest centimeter. Determine the area and perimeter.

17.   Measure and label the sides of this rectangle to the nearest centimeter. Determine the area and perimeter.

18.   To convert from degrees Fahrenheit to degrees Celsius, you use the formula $C = \frac{5}{9}(F - 32)$. What would be the Celsius equivalent of the following temperatures?

a)   32°F

b)   200°F

c)   300°F

19.   To find the hypotenuse of a right triangle with one side of 4 meters and the other side unknown, you would use the formula $H = \sqrt{16 + x^2}$. Find the hypotenuse for right triangles with unknown sides as below:

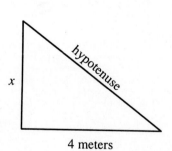

a)   3 meters

b)   6 meters

c)   10 meters.

20.   The area and perimeter of a square with side x are given by $A = x^2$ and $P = 4x$. Find the area and perimeter of squares with sides of these lengths:
a)   4 cm
b) $12\frac{1}{4}$ in

c)   22.5 m
d)   45 yd

21.   The surface area of a sphere with radius r is given by $S = 4\pi r^2$. Find the surface area of spheres with radii as follows:
a)   4 cm
b) $3\frac{1}{4}$ in
c)   7.5 meters

TIP: Use the π key on the TI-83. To use π, press 2nd ^.

22.   The volume of a sphere with radius r is given by $V = \frac{4}{3}\pi r^3$. Find the volume of spheres with radii as follows:
a)   4 cm
b) $3\frac{1}{4}$ in
c)   7.5 meters

TIP: To cube a number on the TI-83, enter the number and then press ^ 3.

23.   The total resistance in a parallel circuit with one resistance equal to 4 ohms (Ω) and the other resistance unknown is given by the formula $R = \dfrac{4r}{r+4}$. What would be the total resistance for circuits with resistor as follows:
a) $3\,\Omega$
b) $5\frac{1}{2}\,\Omega$
c) $6.5\,\Omega$

C.   Complete the following input-output tables.

24.   Complete the following input-output table using $C = \frac{5}{9}(F - 32)$ to convert from degrees Fahrenheit to degrees Celsius.

°F	°C
−212	
$-66\frac{1}{2}$	
0	
32	
$88\frac{2}{3}$	
212	

25.   Complete the following input-output table using $H = \sqrt{49 + x^2}$ to find the hypotenuse of a right triangle with one side of 7 inches and the other side unknown.

x	H
2 in	
4.5 in	
6 in	

26.   Complete the following input-output table using $H = \sqrt{25 + x^2}$ to find the hypotenuse of a right triangle with one side of 5 inches and the other side unknown.

x	H
7.5 in	
10 in	
11 in	
33 in	

27. Complete this input-outputs table finding the area and perimeter of squares with side x using $A = x^2$ and $P = 4x$.

x	P	A
2 in		
4.5 cm		
$6\frac{1}{3}$ ft		
7.5 m		
100 yd		
333.33 km		

28. Complete the following input-outputs table using the given formulas for surface area and volume of a sphere with radius r: $S = 4\pi r^2$ and $V = \frac{4}{3}\pi r^3$.

r	S	V
2 cm		
$3\frac{1}{2}$ in		
4.75 m		
10 yd		
100 m		

29. The total resistance in a parallel circuit with one resistance equal to 4 ohms and the other resistance unknown is given by the formula $R = \dfrac{4r}{r+4}$. Use this formula to complete the input-output table.

r	1 Ω	1.75 Ω	3 ½ Ω	4 Ω	8 2/3 Ω	100 Ω
R						

30. Complete the following table to compute the simple interest for the given investments. Remember that time must be converted to years.

P	R	T	I = PRT
$300	4%	2 months	
$3000	2.73%	4 years	
$10,000	5%	90 days	
$250,000	9%	5 years	
$4,000,000	9.7%	10 days	

D. Use the given graph to answer the following problems.

31. The following graph is the result of an on-line computer survey. The graph demonstrates the number of WWW users according to the user's age of those responding. Complete the input-output table from the information supplied by the graph.

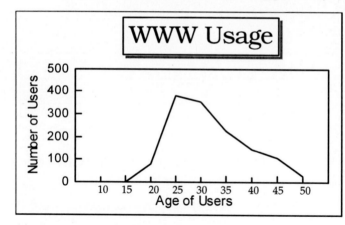

Age of Users	Number of Users
15	
20	
25	
35	
40	
45	

Source: *Georgia Institute of Technology*

32. The following graph is the result of an on-line computer survey. The graph shows the percent of WWW users who sit in front of a computer according to the number of hours they reported. Complete the input-output table from the information supplied by the graph.

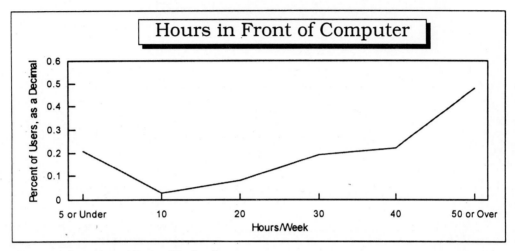

Source: *Georgia Institute of Technology*

Hours/Week	10	20	30	40
Percent of Users				

33. This graph shows the annual U.S. budget deficits as a function of the fiscal year. Answer the questions according to the graph.

    a) Approximately what was the budget deficit for fiscal year 1993?

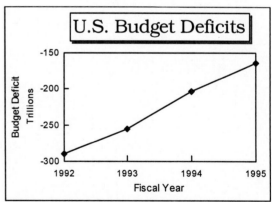

Source: *Office of Management and Budget*

    b) What is the difference between the budget deficits of 1994 and 1995?

    c) What is the total of the budget deficits for the four years shown on the graph?

E.   Complete the following patterns; remember to look for obvious patterns first.

34. $\frac{3}{4}, \frac{6}{8}, \frac{9}{12},$ _____ , _____ , _____

35.   _____ , _____ , _____ , 0, 3, 6, 9

36.   5, 10, 15, _____ , _____ , _____

37.   75, 50, 25, _____ , _____ , _____

38.   25, 36, 49, _____ , _____ , _____

39.   1, 8, 27, _____ , _____ , _____

F.   If possible, complete the following input-output tables.

40.

x	y
0	32
5	23
10	14
15	
20	

41.

Hours worked	Gross Salary
24	$150
28	$175
32	$200
36	
40	

42.

Purchase Price	Sales Tax
$4	$0.10
$14	$0.35
$24	$0.60
$34	
$40	

G.   Complete the following input-output tables. Plot the points on graph
paper and on your calculator. Connect the points to determine the
shape of the graph. Which I-O tables produce graphs that appear to be
lines?

43.

x	$y = 2x + 3$
–2	
–1	
0	
1	
2	

44.

x	$y = 5 - 2x$
–3	
–1	
0	
2	
4	

45.

x	$y = x^2 + 3$
–2	
–1	
0	
1	
2	

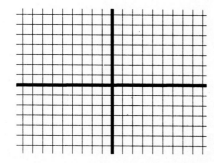

46.

x	$y = \frac{2}{3}x - 3$
–6	
–3	
0	
2	
4	

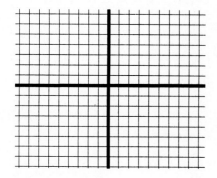

H.   Complete the following input-output table. Plot the points on graph paper and on your calculator. Connect the points to determine the shape of the graph. Does this I-O table produce a graph that appears to be a line?

47.

x	$y = \sqrt{(2x - 4)}$
2	
3	
4	
5	
8	

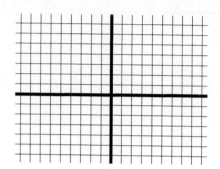

Solving Linear Equations

Upon successful completion of this unit you should be able to:

1. Recognize equations and expressions;

2. Solve linear equations numerically using input-output and calculator tables;

3. Solve linear equations graphically using paper-and-pencil and calculator graphs;

4. Solve linear equations algebraically;

5. Simplify expressions with parentheses using the distributive law;

6. Solve linear equations containing parentheses graphically, numerically using tables, and algebraically; and

7. Set up and solve linear equations from applications.

What is an Equation?

When most people think of algebra, they think of solving equations. One may wonder, however, what is an equation?

In Unit 1, we used variables to write formulas such as $A = LW$ and $y = 2x - 3$. The expressions on both sides of a formula are called variable expressions. Therefore, A, LW, y, and $2x - 3$ are variable expressions of $A = LW$ and $y = 2x - 3$.

> If we place an equal sign between two variable expressions or between a variable expression and a number, we get an equation.

An equation has an equals sign, =, and is in the form

$$Expression1 = Expression2.$$

 Label each of the following as either an expression or an equation.

1. $2x + 3 = 8$

   Using the definition of an equation, $2x + 3 = 8$ is an equation with $2x + 3$ as *expression1* and 8 as *expression2*.

2. $3 - 4x^2 = 2x + 5$

   $3 - 4x^2 = 2x + 5$ is an equation with *expression1* $= 3 - 4x^2$ and *expression2* $= 2x + 5$.

3. $4 - (2x - 7)$

   $4 - (2x - 7)$ is not an equation, because it consists of a single expression.

4. $xy + 4 - 7$

   $xy + 4 - 7$ is an expression.

You Try It

Which of the following are equations?

1. $3x + 7 = 44$

2. $3 - (2x - 7)$

3. $x + 33 = 2x + 7$

4. $4(2x + 3) = 3 - (2x - 7)$

Solving Equations

> To solve an equation means to find the input that yields a specific output.

But I Gotta Get a B!

We solve equations to find the answer to a problem in everyday life or at work. Greg needs help in allocating his study time during the rest of the term. He is taking a mathematics class that requires 650 total points to earn a B. He has accumulated 476 points to date and has two 100 point tests remaining. What does he need to average on the two tests to earn a B?

Let T represent the average grade on his next two tests. Then $2T$ would represent the points that Greg would earn on the two tests and $476 + 2T$ would represent the total points that he would have after the two tests.

Greg needs 650 points for a B. Therefore, he should solve $476 + 2T = 650$ to find what he needs to average on the next two tests. The output is 650, and the average test score is the input.

In this unit, we will find the solution to this and other equations.

Solving Equations Using Tables

CoolNet offers Internet connection for students for a flat monthly fee of $3 with a $2 per hour on-line fee. If this month you budgeted $8, to determine how many hours you can stay on-line, you might solve the equation $2h + 3 = 8$, where h is the number of hours on-line.

Here is an input-output table:

h	$2h + 3$
0	3
1	5
2	7
3	9
4	11

Are any of the values of $2h + 3$ equal to 8?

No, but the output for 2 is 7 and the output for 3 is 9. Therefore, the input that results in 8 would be between 2 and 3. Why?

Let's try increments of 0.1 instead of 1. An **increment** is the change from one input value to the next.

h	$2h + 3$
2.3	7.6
2.4	7.8
2.5	8
2.6	8.2
2.7	8.4
2.8	8.6

Note that an input of 2.5 gives an output of 8. Therefore, $h = 2.5$ is the solution of the equation $2h + 3 = 8$. You can stay on line 2.5 hours and be within your budget of $8.

You Try It

Suppose the length of a rectangle is 4 less than 3 times its width, x. We can then express the length as $3x - 4$. If we know that the length is 6 meters, use input-output tables to solve $3x - 4 = 6$.

1. Complete the following table to determine a range of values for the solution:

x	$3x - 4$
0	
1	
2	
3	
4	

2. The solution is between $x =$ _____ and $x =$ _____. Why?

3. Now use a second input-output table to find the answer to the nearest tenth:

x	$3x - 4$

4. The solution to the nearest tenth is $x =$ _____.
    Therefore, the width of the rectangle to the nearest tenth of a meter is _____.

Return to Greg's Equation

We will now use input-output tables to solve $476 + 2T = 650$, the equation we developed to represent the situation with Greg's grade.

Since the numbers involved are large, let's start with a table whose inputs have increments of 10:

T	476 + 2T
70	616
80	636
90	656
100	676

The input column contains values of T representing the grade he will average on the next two tests. The output column contains corresponding output values $476 + 2T$ representing the total points that he will earn for the course.

Remember that Greg needs 650 points to earn a B. Notice that the row with input $T = 80$ gives 636 total points while the row with input $T = 90$ gives an output of 656 total points. Therefore, the answer for the average grade is between $T = 80$ and $T = 90$ and seems to be closer to $T = 90$. Why?

Change the increment in the table to 1 point rather than 10 points. Complete the table, and use it to answer the question following the table.

T	476 + 2T
85	
86	
87	
88	
89	
90	

Greg will reach his goal if he averages _____ on the next two tests.

You Try It

Tara is a friend of Greg's, who is enrolled in a different section of the same course. She has three tests to complete instead of two. She has accumulated 387 points thus far and also needs 650 total points to get a B.

1. Complete the following input-output table, and use it as the first step to find the solution to $387 + 3T = 650$.

T	$387 + 3T$
75	
80	
85	
90	
95	
100	

2. The solution would be between $T =$ _____ and $T =$ _____.

3. Use values of T that vary by 1 point for this second table.

T	$387 + 3T$

4. The solution would be $T =$ _____ when rounded to the nearest point. Tara needs to average _____ on the next three tests to earn a B.

A Little Competition

As before, Greg has accumulated 476 points to date and has two 100-point tests left in his mathematics class. Tara has 387 points to date and has three 100 point exams remaining. Tara would like to accumulate the same number of total points as Greg. What grade must each of them average to have the same total points after all their tests?

If T represents the average grade they need on the remaining tests, Greg would have $476 + 2T$ points and Tara $387 + 3T$ points. To answer the question, we need to solve the equation

$$476 + 2T \text{ points} = 387 + 3T \text{ points.}$$

Here is an input-outputs table with a column for each side of the equation.

Average Needed	Greg's Points	Tara's Points
T	$476 + 2T$	$387 + 3T$
80	636	627
85	646	642
90	656	657
95	666	672

Greg has more points than Tara for the rows with average test scores of 80 and 85. Tara has more points for the rows with average test scores of 90 and 95. Therefore, they should have the same number of points somewhere between averages of 85 and 90.

 Complete the following input-outputs table to estimate the average grade needed to accumulate the same number of total points.

Average Needed	Greg's Points	Tara's Points
T	$476 + 2T$	$387 + 3T$
86		
87		
88		
89		
90		

They both need to average _____ to have the same total points of _____.

You Try It

Construct tables to solve $14x - 5 = 3x + 20$ to the nearest 0.1.

Solving Equations Using Tables on the Calculator

There are several ways to use the graphing calculator to find the solutions of an equation. In this section, we will use the table feature to find solutions in a way similar to the input-output table method of the last section.

Let's return to Greg's problem of solving $476 + 2T = 650$. Look at the first input-output table for $476 + 2T$. Note that the first value of the table was $T = 70$ and that the values of T increased by 10 each time.

We will now enter the formula for Greg's total points in the calculator and set up a table to help solve the equation $476 + 2T = 650$.

For the calculator, we must use x instead of T. Most mathematics classes use x as a generic variable while we chose T for this problem to remind us that it represents the average test score needed.

Turn your calculator on, and press the [Y=] key.

Enter $476 + 2x$ in **Y1**. Your calculator screen should look like the one to the right.

```
Plot1  Plot2  Plot3
\Y1=476+2X■
\Y2=
\Y3=
\Y4=
\Y5=
\Y6=
\Y7=
```

Press [2nd][WINDOW] to set up the table.

Remember that the first value of the test average for the input-output table in the previous section was $T = 70$ and that the values of T increased by 10.

Enter 70 for the value of **TblStart** and 10 as **ΔTbl**. Delta (Δ) is a Greek letter used in mathematics to indicate change. Mathematically, this change is called an increment.

```
TABLE SETUP
 TblStart=70
 ΔTbl=10
Indpnt: Auto  Ask
Depend: Auto  Ask
```

Now press [2nd][GRAPH] to view the table.

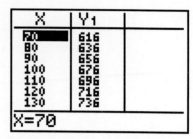

Remember the goal is for the test average **Y1** to equal 650 points for a B. Notice that values of **Y1** are closest to 650 between the values of $x = 80$ and $x = 90$, so the solution to the problem is between those values.

Let's create another table beginning with $x = 85$ and an increment of 1 point.

Press [2nd][WINDOW] to set up the table. Enter 85 for the value of **TblStart** and 1 as △**Tbl**.

Your screen should look like the one below.

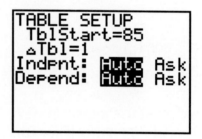

Now press [2nd][GRAPH] to view the table and press the down arrow until **Y1** is 650.

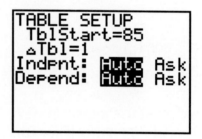

Notice that the value of $x = 87$ has a **Y1** value of 650. Therefore, Greg must score an average of 87 points on the next two tests to earn a B.

Note: There is a calculator appendix at the end of this book. Please refer to it if you need more details on how to use the calculator. Ask your instructor if you have questions not answered by the text or appendix.

You Try It

Construct tables on your calculator to find the solution to the equation $2x + 3 = 8$. Use increments (Δ**Tbl**) of 1 for your first table and 0.1 for the second table. Use a beginning value of $x = -5$ for the first table. Clear the current equation in **Y1**, and then enter $2x + 3$ in **Y1**.

Record at least four rows from each of your tables, and use a complete sentence to explain your answer.

The Solution of "A Little Competition" Using the Calculator

Remember that Tara has one more test to take than Greg, and she wants to have the same total points at the end of the semester. We found that she needed to solve the following equation to answer her question:

$$476 + 2T = 387 + 3T$$

To solve this on the TI-83, we would enter *expression1*, $476 + 2T$ in **Y1**. Then enter *expression2*, $387 + 3T$ in **Y2**. Use x instead of T on the calculator.

```
Plot1 Plot2 Plot3
\Y1■476+2X
\Y2■387+3X
\Y3=
\Y4=
\Y5=
\Y6=
\Y7=
```

Press 2nd WINDOW to set up the table. Enter 80 for the value of **TblStart** and enter 5 as Δ**Tbl**.

```
TABLE SETUP
 TblStart=80
 ΔTbl=5■
Indpnt:  Auto  Ask
Depend:  Auto  Ask
```

Now press 2nd GRAPH to view the table.

X	Y1	Y2
80	636	627
85	646	642
90	656	657
95	666	672
100	676	687
105	686	702
110	696	717

X=80

Greg has more points than Tara for the rows with averages of 80 and 85, but Tara has more points for each row with an average of 90 or greater. Therefore, Greg and Tara should have the same number of total points between averages of 85 and 90.

Use 85 for **TblStart** and 1 as Δ**Tbl** to create another table.

```
TABLE SETUP
 TblStart=85
 △Tbl=1
Indpnt: Auto  Ask
Depend: Auto  Ask
```

Use the table to complete this sentence. Tara and Greg would need averages of _____ to end up with total points of _____.

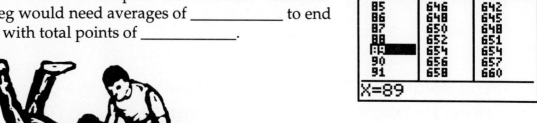

X	Y₁	Y₂
85	646	642
86	648	645
87	650	648
88	652	651
89	654	654
90	656	657
91	658	660

X=89

You Try It

Use the calculator's table feature to solve the equation

$$14x - 5 = 3x + 20.$$

Write the values that you use for **TblStart** and Δ**Tbl** for each table. Copy several rows of the table to show how you reached the solution or how you made the decision to begin the next table. Use a complete sentence to explain your answer.

Solving Equations Graphically

 ▶ Consider the equation $3x - 4 = 5$. To solve $3x - 4 = 5$ means we want to find the input value of x that yields the output value of 5.

Another way to solve the equation is to find the point of intersection of the graphs of the two sides of $3x - 4 = 5$. We plotted the line $y = 3x - 4$ from an I-O table in Unit 1. If we also plot $y = 5$, we observe that it is a horizontal line crossing the y-axis at 5.

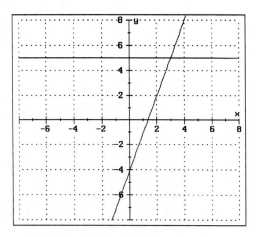

The lines cross at $x = 3$ and $y = 5$. The solution of the equation $3x - 4 = 5$ is the x-value where the y-value of each line is the same. The input value $x = 3$ yields the output value $y = 5$.

 1. Use the graph to solve $3x - 4 = 2$. The two equations graphed are $y = 3x - 4$ and $y = 2$, which is a horizontal line.

The point of intersection is $(x, y) = $ _____.

The solution is _____.

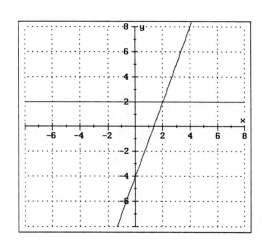

2. Add a horizontal line at $y = -1$, and use the graph to solve $3x - 4 = -1$.

The point of intersection is $(x, y) = $ _____.

The solution is _____.

You Try It

1. Look at the graph of $y = 2x + 3$.

    Use a ruler or straight edge to add horizontal lines at $y = -5$ and $y = 4$, and complete the table.

    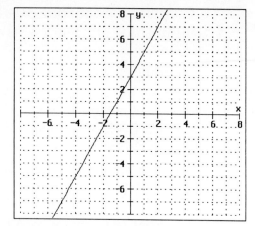

Equation	Point of Intersection	Solution
$2x + 3 = -5$		
$2x + 3 = 4$		

2. The graphs of $y = 3x - 7$, $y = -4$, $y = 0$, $y = 4$, and $y = 8$ are at the right.

    Use the graph to complete the table below.

    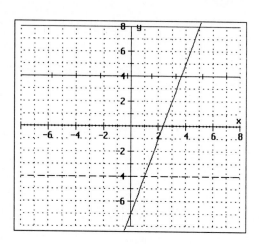

Equation	Point of Intersection	Solution
$3x - 7 = -4$		
$3x - 7 = 0$		
$3x - 7 = 4$		
$3x - 7 = 8$		

An Equation Where You Draw the Graphs

Complete the input-output table, graph the points, and sketch the graph of $y = \frac{1}{2}x + 2$.

x	y
−4	
−2	
0	
2	
4	

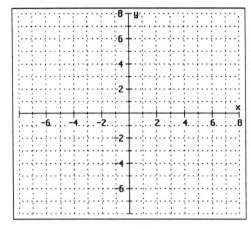

Add horizontal lines, and use the graphs to solve the six equations which follow.

1. $\frac{1}{2}x + 2 = 6$

2. $\frac{1}{2}x + 2 = 3$

3. $\frac{1}{2}x + 2 = 1$

4. $\frac{1}{2}x + 2 = 0$

5. $\frac{1}{2}x + 2 = -1$

6. $\frac{1}{2}x + 2 = -2$

Solving Equations Graphically on the Calculator

We will now use the calculator to solve equations graphically. We do this by graphing two lines and finding their point of intersection as we did on paper in the previous activity. The first line uses the expression from the left side of the equation and the second uses the expression on the right side.

 Solve $2x + 3 = 5$.

We use the two lines $y = 2x + 3$ and $y = 5$ to solve $2x + 3 = 5$.

Turn your calculator on, and press the Y= key. Clear any expression currently in the calculator.

Enter $2x + 3$ in **Y1** and 5 in **Y2**.

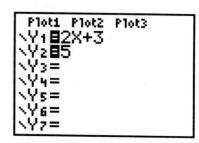

Setting the Window

Next we need to set the graphing window. This is similar to deciding how large a piece of graph paper to use. Complete the input-outputs table to help you decide what window to use.

x	$2x + 3$	5
–5		
0		
5		

Based on the table, we want to use a piece of graph paper with x-values from –5 to 5 and y-values from –7 to 13. We write the window as $[-5, 5]_x$ and $[-7, 13]_y$. Note that –5 and 5 were arbitrary choices for x. The window for this graph would change if you chose other values of x.

Press the WINDOW key.

Enter –5 for **Xmin**, 5 for **Xmax**, –7 for **Ymin**, and 13 for **Ymax**.

Press [GRAPH] to see the lines $y = 2x + 3$ and $y = 5$.

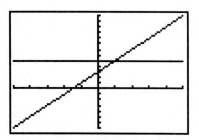

Now press [TRACE] to find the point of intersection.

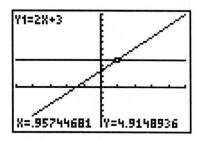

Notice the blinking cursor on your screen. Use the right arrow key to move the cursor to the point where the two lines cross.

The graphs cross at approximately (0.96, 4.9). The solution to $2x + 3 = 5$ is $x \approx 0.96$. This is not the exact solution since the y-value is not 5. How can we change the window to improve the approximation?

The screen of the TI-82 or TI-83 has a width of 94 pixels. The pixels on the calculator screen determine the x-values as you trace. We could use windows like $[0, 94]_x$, $[0, 9.4]_x$ or any other $[$**Xmin, Xmax**$]_x$, where **Xmax – Xmin** is divisible by 94. For the window found by pressing [ZOOM][4], $[-4.7, 4.7]_x$, $4.7 - (-4.7) = 9.4$, which is one-tenth of 94. Therefore, the x-values change by tenths. The y-value at each pixel is a function of the x-value. The calculator substitutes the traced x-value into **Y1**.

Press the [WINDOW] key.

Using the same y-values, enter –4.7 for **Xmin** and 4.7 for **Xmax**.

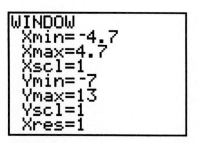

Press [GRAPH] and [TRACE]. Use the arrow keys to locate the point of intersection. The point of intersection is $(x, y) = ($ _____ , _____ $)$. The solution to $2x + 3 = 5$ is $x = 1$.

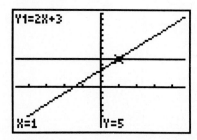

Solve: $2x + 3 = -7$

Replace the 5 in **Y2** with –7 to solve the equation, $2x + 3 = -7$. Use the same window as before.

The two lines do not appear to intersect on this window. Why?

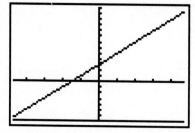

Change the *y*-values of the window to $[-13, 13]_x$. The lines still do not intersect so we need to change the *x*-values on the axes. If we add an integer to both **Xmax** and **Xmin**, we will shift the viewing window, but maintain the difference of 9.4 so the *x*-values will still be tenths.

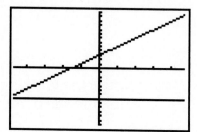

To shift the viewing window to the left, we subtracted 2 from both **Xmax** and **Xmin**.

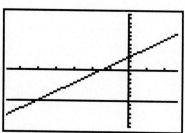

Press TRACE. Use the left arrow key to move the cursor to the intersection point.

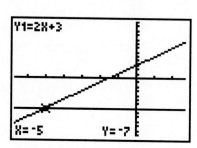

The solution is $x = -5$ since the graphs cross at $(x, y) = (-5, -7)$.

You Try It

1. Replace the value in **Y2**, change the window, and trace to complete the following table.

Equation	Window	Point of Intersection	Solution
$2x + 3 = -3$			
$2x + 3 = 0$			
$2x + 3 = 3$			
$2x + 3 = 4$			

2. How can you confirm the solution you obtained graphically?

Return to a Little Competition

Remember that we solved the equation $476 + 2T = 387 + 3T$ to determine the average Greg and Tara needed to have the same number of total points.

To solve this equation graphically, we'll complete the following process:

1.   Enter $476 + 2x$ in **Y1** and $387 + 3x$ in **Y2**.

2.   Choose an appropriate window for the graph.

3.   Graph the lines.

4.   Trace to find the point of intersection.

If we begin with a window of $[-5, 183]_x$ and $[-50, 1000]_y$ and follow the process above, we get these screens and the graphs on the next page.

```
Plot1 Plot2 Plot3
\Y1■476+2X
\Y2■387+3X
\Y3=
\Y4=
\Y5=
\Y6=
\Y7=
```

```
WINDOW
 Xmin=-5
 Xmax=183
 Xscl=10
 Ymin=-50
 Ymax=1000
 Yscl=50
 Xres=1
```

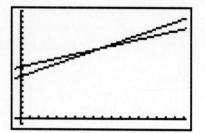

Press TRACE , and use the arrow keys to locate the point of intersection, if possible. We used the 94 pixels multiplied by 2, so when you trace, you will notice that the increment between the *x*-values is 2. If **Xmin** were −10 and **Xmax** were 178, the cursor would have skipped the solution of 89. Why?

Because of our careful selection of the window, tracing shows the point of intersection to be (89, 654). Often you will need to adjust the window to locate the point of intersection.

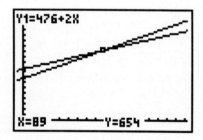

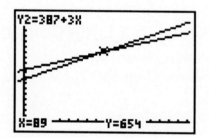

When you press the up arrow, the tracing function moves from one graph to the next. The calculator substitutes the same *x*-value into the next function. When the *y*-values are constant, you have found the point of intersection. As before, Greg and Tara each need to average 89 on the remaining tests to have the same number of total points.

You Try It

Use calculator graphs to solve the following equations.

	Equation	Window	Point of Intersection	Solution
1.	$5x + 4 = -6$			
2.	$3x - 1 = 2x + 5$			
3.	$\frac{2}{3}x + 8 = \frac{-3}{5}x - 11$			

Solving Equations Algebraically

It is often easier, quicker, and more convenient to solve equations algebraically. We do this by changing the given equation to an equivalent one of the form

$$Variable = Constant,$$

where the constant is the solution of the original equation. When we substitute the constant into the original equation, the result is a true statement.

When we solve equations, we create equivalent equations by

> 1. Adding the same number to both sides of the equation;
>
> 2. Subtracting the same number from both sides of the equation;
>
> 3. Multiplying both sides of the equation by the same non-zero number;
>
> 4. Dividing both sides of the equation by the same non-zero number.

Why do **3** and **4** above say the number must not be zero?

You produce an equivalent equation when you perform the same operation on both sides of it.

1.   Solve $x - 3 = 2$ by adding 3 to both sides.

$$\begin{aligned} x - 3 &= 2 \\ x - 3 + 3 &= 2 + 3 \\ x &= 5 \end{aligned}$$

2.   Solve $x + 3 = 8$ by subtracting 3 from both sides.

$$\begin{aligned} x + 3 &= 8 \\ x + 3 - 3 &= 8 - 3 \\ x &= 5 \end{aligned}$$

3.   Solve $\dfrac{x}{3} = \dfrac{5}{3}$ by multiplying both sides by 3.

$$\frac{x}{3} = \frac{5}{3}$$

$$3\left(\frac{x}{3}\right) = 3\left(\frac{5}{3}\right)$$

$$x = 5$$

4.   Solve $3x = 15$ by dividing both sides by 3:

$$3x = 15$$

$$\frac{3x}{3} = \frac{15}{3}$$

$$x = 5$$

In solving an equation, we change to an equivalent equation in the form *variable = constant*, where the constant is the solution. We do this using opposites. In other words, we reverse or undo addition by subtraction and multiplication by division.

You Try It

Solve for x.

1. $x - 3 = -2$ **2.** $2 + x = -5$

3. $\dfrac{x}{-6} = 7$ **4.** $35 = -7x$

Greg's Equation

How would we solve Greg's equation algebraically? Remember the equation is

$$476 + 2T \text{ points} = 650 \text{ points.}$$

We need to solve for T. To solve for T, we need the $2T$ on one side and the numbers on the other side.

We can do this by first subtracting 476 from both sides:

$$\begin{aligned}
476 + 2T &= 650 \\
476 + 2T - 476 &= 650 - 476 \\
2T &= 174
\end{aligned}$$

We now can solve for T by dividing both sides by 2:

$$\begin{aligned}
2T &= 174 \\
\frac{2T}{2} &= \frac{174}{2} \\
T &= 87
\end{aligned}$$

This gives a solution of $T = 87$. Therefore, Greg needs to average 87 points to earn a B in the course.

Tara's Equation

Let's solve Tara's problem algebraically: $387 + 3T = 650$.

1.   Subtract 387 from both sides.

2.   Divide both sides by 3.

3.   Therefore, Tara needs to average _____ on her last three tests to earn a B.

You Try It

Solve the following equations algebraically and check.

1. $2x + 3 = 9$ **2.** $3y - 5 = 6$

3. $4z + 80 = 14$ **4.** $5a - 3 = -11$

5. $4t - 17 = -330$

A Little More Competition

Remember that Tara has one more test to take than Greg and wants to have the same number of total points at the end of the semester. To answer her question, Tara needs to solve

$$476 + 2T \text{ points} = 387 + 3T \text{ points.}$$

Looking at this equation, there is a T on each side. We'll begin by subtracting $3T$ from both sides in order to eventually have *variable = constant*:

$$476 + 2T = 387 + 3T$$
$$476 + 2T - 3T = 387 + 3T - 3T$$

Unit 2 Problems for Practice

A.   Complete the following:

1. $\dfrac{2}{3} - \dfrac{1}{4} =$ 2. $(-2.5)^2 =$

3.   Determine the area and perimeter of the rectangle shown here.

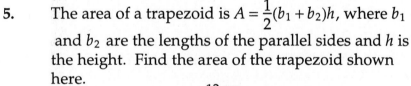

4.   The volume of a cylinder is $V = \pi r^2 h$. If a soft drink can is 12 cm high with a radius of 3.1 cm, what is the volume in cubic centimeters?

5.   The area of a trapezoid is $A = \dfrac{1}{2}(b_1 + b_2)h$, where b_1 and b_2 are the lengths of the parallel sides and h is the height. Find the area of the trapezoid shown here.

12 cm

5 cm

15 cm

6.   The circumference of a circle is $C = 2\pi r$, where r is the radius. Complete the following I-O Table, rounding to the nearest hundredth.

r	2	4	6	8	10
C					

7.   The mean (average) of two numbers can be determined from the formula $A = \frac{a+b}{2}$. Complete the following I-O table for two tests, if you already know that on the first test you scored 92 points.

b	100	95	90	85	80	75
A						

8.   Complete the following I-O table, where $y = -2x + 3$, and plot points on a graph.

x	−10	−5	0	5	10	15
y						

9.   Write the next three terms of the following sequences by looking for an obvious pattern:

a)   0, −4, −8, −12, …

b)   1, 4, 7, 10, …

10.   On many brands of running shoes, sizes are listed for U.S. sizes, U.K. sizes, and European sizes. Plot the following results for pairs of shoe sizes, where the horizontal axis represents U.S. sizes and the vertical axis represents European sizes. Do the points appear to lie in a line?

U.S. Sizes	Eur. Sizes
9.5	42.75
10	43.5
4.5	36.25
6	38.25

11.   The graph below shows the median earnings of females with Associate's degrees as a function of the year. Complete the input-output table.

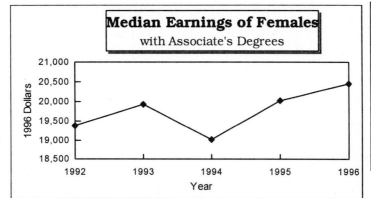

Year	1996 Dollars
	19,000
	20,460
1992	
1993	
1994	

Source: *U.S. Bureau of the Census*

12.   This graph shows the median earnings of females with some college, but without an Associate's degree as a function of the year. Complete the input-output table.

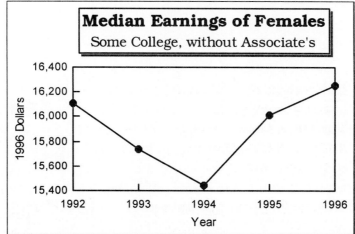

Year	1996 Dollars
1993	
1994	
1996	
	16,000
	15,440

Source: *U.S. Bureau of the Census*

B.   Which of the following are equations? Identify *expression1* and *expression2* in each equation.

13. $2x - 5 = 4$ 14. $xy + 2x + 3$

15. $2(x + 7) - 3(x - 5)$ 16. $3 = 2x - 7$

17. $4(2x + 3) = 3 - (2x - 7)$ 18. $xy = 2x + 3$

19. $5x^2 = 80$

20. $5y + 3 = 4 - (y - 8)$

C.   Solve the following equations by completing input-output tables:

21. $3x + 1 = 10$

x	

22. $4 + 6n = 7$

n	
0.6	

23. $75 - 11b = 113.5$

24. $2x + 7 = 5x + 1$

25. $2 - 7x = 3x + 12$

26.   Luke wants to buy a pair of German sandals from a catalog. Only the European size is available, but Luke knows that the U.S. size, u, is related to the European size, E, by the equation $u = 0.76E - 23.2$. If Luke wears a U.S. Size 8, what size should he order? Use an input-output table to answer the question.

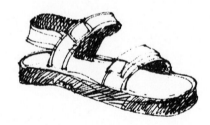

D.   Construct tables on your calculator to solve these equations. Type the variable expression into **Y1**. Give at least three rows of your table as well as the answer. [Tip: Start with **TblStart** = 0 and Δ**Tbl** = 1.]

27. $5 + 2x = 9$

28. $5 - a = -8$

29. $2d + 40 = 59$

30. $75 - 11b = 113.5$

31. $45g - 17 = 532$

32.   A local factory job pays \$290 a week for summer employment with a \$2.75 bonus per unit assembled. Wanting to make \$350 before taxes, solve the equation $350 = 290 + 2.75u$ to determine the least number of units you must assemble.

E.   Solve the given equations. Draw horizontal lines as needed and locate the points of intersection to solve.

33.   **a)** $3x - 4 = 5$

        b) $3x - 4 = 2$

        c) $3x - 4 = -1$

        d) $3x - 4 = -7$

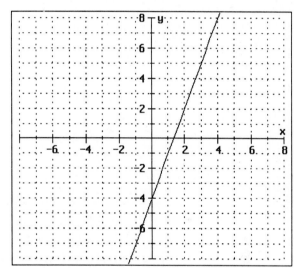

34.   **a)** $2x + 3 = 5$

        b) $2x + 3 = 2$

        c) $2x + 3 = 0$

        d) $2x + 3 = -3$

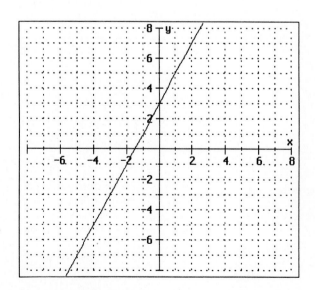

35.   **a)** $\frac{2}{3}x + 1 = 6$

b) $\frac{2}{3}x + 1 = 3$

c) $\frac{2}{3}x + 1 = 0$

d) $\frac{2}{3}x + 1 = -3$

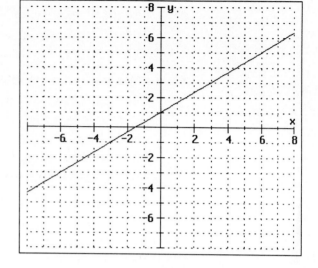

36.   **a)** $\frac{3}{4}x - 3 = 2$

b) $\frac{3}{4}x - 3 = 1$

c) $\frac{3}{4}x - 3 = -2$

d) $\frac{3}{4}x - 3 = -7$

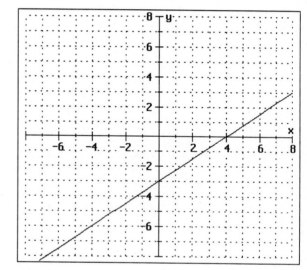

37.   **a)** $5 - x^2 = 3$

b) $5 - x^2 = 1$

c) $5 - x^2 = -4$

d) $5 - x^2 = -7$

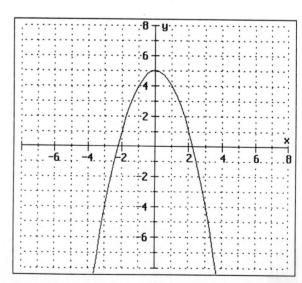

F.   a)   Graph the line $y = expression1$ using an I-O table;
   b)   Graph the line $y = expression2$ by sketching a horizontal line through the point $(0, expression2)$;
   c)   Find the point of intersection; and
   d)   Record the solution of the original equation.

38. $3x + 5 = 8$

x					
y					

39. $2x - 5 = 4$

x					
y					

40. $\frac{1}{2}x + 3 = 7$

x					
y					

41. $\frac{2}{3}x - 5 = 11$

x					$=8$
y					

42. $-2x + 15 = -1$

x					$=8$
y					

G. Solve graphically using your calculator. The first one was started for you as an example.

		Y1 =	**Y2 =**	**Graphing Window**	**Solution**
Example:	$x + 5 = 7$	$x + 5$	7	$[-5, 4.2]_x$ $[-5, 12]_y$	
43.	$-2x - 5 = 3$	$2x - 5$	3		-4
44.	$12 - b = 17$				-5
45.	$45g - 17 = 532$				12.2
46.	$2 - 7x = 3x + 12$				-1

H. Solve the following equations algebraically:

47. $3x + 4 = 6$

48. $-2y - 3 = -6$

49. $2z + 1 = 4$

50. $5 + 3x = 7 - 2x$

51. $x - 7 = 4 + 3x$

52. $5y - 11 = 2 - y$

I. Solve the following equations algebraically.

53. $4 + 8a = -10$

$A = 1.8$ $8a = -10 + 4$

$8a = -14$

54. $9 - 2z = z + 5$

$z = 1.3$

55. $4b - 17 = 19 - b$

$b = 7.2$

56. $2x + 13 = 5 - x$

$x = 2.7$

J.   Simplify the following expressions:

57. $2(3x-4)$

58. $3(2x-3)+5(4-2x)$

59. $3(2+5y)$

60. $-2(6-3z)+4(3z-5)$

61. $5(-3a+4)$

62. $-2(1+5x)$

63. $-6(7-2b)$

64. $8(3g+2h)+3(2g-5h)$

65. $-2(5-3x)$

66. $-2(-7y+6x)-3(4x-5y)$

K.   Solve the following equations graphically, numerically, and algebraically:

67. $2(5x-2)=3(1-x)$
$x=.5$

68. $5(2x+4)=2(3-5x)$
$x=-1.7$

69. $7(x-2)=4(1+3x)$
$x=-3.7$

70. $-2(1-x)=5(2x+3)$
$x=2.1$

71. $3(4-2x)=-3(x-5)$

72. $6(x+3)=-2(x+4)$

73. $7(1+x)=-11(5-x)$

74. $-(x+4)=3(5-2x)$

75. $3(2x-6)=2(1+7x)$

76. $-2(4x+3)=-(5-x)$

L.   Work the following group problems.

77.   If the perimeter of a rectangle is 50 cm and the length is 15 cm, what is the width?

78.   Jo has 2 quizzes left in a class and has accumulated 62 total quiz points to date. She needs a total of 80 points to earn a B quiz average. Write an equation that uses q to represent the quiz average needed, and solve it.

79.   Write an equation that expresses the area for the following situation:

Maria wants a pool with a 2-foot wide sidewalk surrounding it. She wants the pool to be twice as long as it is wide. Let x represent the width of the pool. Write the equation for the area of the sidewalk.

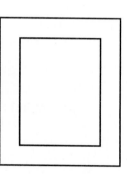

80.   Which of the following expressions are not equivalent? Why not?

a) $-2(x + 5) - 2$ **b)** $-2x + 8$

c) $-2(x + 6)$ **d)** $-12 - 2x$

Simplifying Expressions

Upon successful completion of this unit you should be able to:

1. Simplify variable expressions involving addition, subtraction, multiplication, and division;

2. Simplify variable expressions involving integer exponents;

3. Calculate areas and perimeters of geometric figures with dimensions written as variable expressions;

4. Use scientific notation to estimate answers to problems; and

5. Use scientific notation to solve problems involving very large or very small numbers.

Geometric Figures

Note: Figures are not necessarily drawn to scale.

1.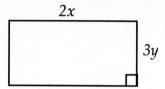

2.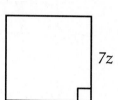

3.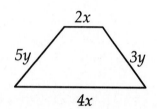

4.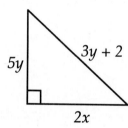

5.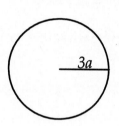

6.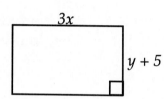

7.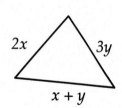

8.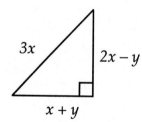

9.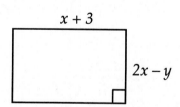

10.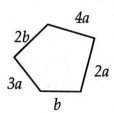

Introduction to Variable Expressions

The geometric figures on the opposite page have dimensions written as variable expressions. Later we will find the perimeter of these figures.

But first, let's examine the parts of a variable expression. One variable expression with two terms is $7x + 8$.

The expression $7x + 8$ has two terms separated by a plus sign. The first term is $7x$. Seven is the coefficient while x is the variable part. The second term is the constant 8.

 The variable expression $8x - y + 3xy - 5$ has four terms separated by plus and minus signs. List the four terms and name the coefficient and variable parts.

1.   The first term is $8x$. Name the coefficient and variable parts.

2.   The second term is $-y$. The coefficient is -1 while y is the variable part. The third term is $3xy$. Three is the coefficient while xy is the variable part. The constant -5 is the fourth term.

Later in the unit, we will examine variable expressions in more detail. Now let's find the perimeter of each figure on the preceding page. This will require us to add coefficients of like terms. Like terms are terms with exactly the same variable part.

Recall from Unit 1 that the perimeter of a geometric figure is the distance around the figure. We find the perimeter by adding the lengths of its sides. We often use shortcuts for the process of adding repeatedly. For example, for a square we could add side + side + side + side or we could just multiply 4 times the length of its side.

> Rectangle: $P = 2 \cdot length + 2 \cdot width$
> Square: $P = 4 \cdot side$ $P = 2(L) + 2(W)$
> Triangle: $P = side + side + side$
> Circle: $C = \pi \cdot diameter$

Figure 1   This rectangle has length of $2x$ and width of $3y$. The perimeter is $P = 2(2x) + 2(3y)$, so $P = 4x + 6y$. Both x and y represent unknown numbers, so $4x + 6y$ has no like terms and is the simplified form of the answer.

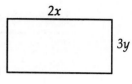

Figure 2   This square has a side of $7z$. The perimeter is $P = 4(7z) = 28z$, which is in simplest form.

Figure 3   This quadrilateral has sides of $4x$, $3y$, $2x$, and $5y$. The perimeter is $P = 4x + 3y + 2x + 5y$. We can combine like terms so $P = 6x + 8y$.

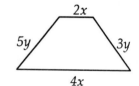

The terms $4x$ and $2x$ are like terms with the same variable part x. We add the coefficients of the x terms, $4x + 2x = 6x$. Similarly, $3y$ and $5y$ are like terms with the same variable part y. The sum of the coefficients of the y terms is $3 + 5 = 8$, so $3y + 5y = 8y$.

> When an algebraic expression contains like terms, we combine the like terms by adding coefficients.

Figure 4   This triangle has sides of $2x$, $5y$, and $3y + 2$. The perimeter is $P = 2x + 5y + 3y + 2$ and simplified $P = 2x + 8y + 2$.

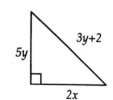

The expression has one x term, two y terms, and the constant 2. We combine the coefficients of the y terms, so $5y + 3y = 8y$. There are no more like terms, and the simplified answer is $P = 2x + 8y + 2$.

You Try It

Determine the circumference for Figure 5 and the perimeter for Figures 6–10 on page 92.

Simplification of Variable Expressions

Earlier, we found the perimeter of geometric figures with sides written as variable expressions. To do this, we added coefficients of like terms.

What is a like term? In the previous unit, we said that like terms have exactly the same variable part. It is easy to agree that x is like x, that x is unlike y, and that a is unlike b. What about xy and yx? Are they like terms?

Remembering that x, y, a, and b each represents an unknown number, we might look at the situation as follows: Suppose $x = 3$ and $y = 4$, what does xy equal? What does yx equal? In each case, because of the commutative property of multiplication, the order of multiplication does not change the result.

> **Like terms** have the same combination of letters. If any of the letters have exponents, then the terms are like terms only if the exponents are the same.

Expressions are sometimes classified according to how many terms they have. Terms are separated by the operation of addition. Subtraction is the addition of the opposite, so $x - y$ has two terms, x and $-y$. Examine the following table.

Expression	Number of Terms	Classification
$3x$	1	Monomial
$3x + 4y$	2	Binomial
$3x - 4y + 5$	3	Trinomial
$2x - 5y + xy - 7$	4	Polynomial
$7 - 3x - 4y - 8xy + x^2 - y^2$	6	Polynomial

 Classify the following expressions. Do they contain like terms? Why?

1. $4xy + 6yx$

    This binomial has like terms because both contain the variables x and y. We say that the expression xy is equivalent to yx because of the commutative property of multiplication.

2. $3bd + 5bc$

This binomial does not contain like terms because the terms have different variables. The first term has b and d while the second has b and c.

3. $7x^2y + 6yx^2 - x$

This trinomial has like terms because the first two terms contain variable expressions equivalent to x^2y.

4. $7 - 4x + 5y - 7x^2y - 6xy^2$

This polynomial does not contain like terms. Look at the last two terms. They are not equivalent. One term contains x^2 and y while the other contains x and y^2.

5. $4abc + 6bc$

This binomial does not contain like terms because the first term is the product of a, b, and c while the second is the product of only b and c.

You Try It

List the number of terms, and classify the following expressions. Determine which contain like terms. Why?

1. $4xy + 6yz - 5$

2. $3ab - 5ab$

3. $7x^2z - 6x^2z + x - y - 3$

4. $6 - 7xy + 6xy^2$

5. $a - b + 8ab - 12abc$

6. $3 - 4xyz - 3xzy$

Later in this unit, some multiplication problems will require us to combine like terms. First, let's work with simplifying expressions with exponents.

Integer Exponents

We write $x \cdot x = x^2$. Similarly, $x \cdot x \cdot x = x^3$ and $x \cdot x \cdot x \cdot x = x^4$. Exponents are a shorthand way of writing repeated multiplication.

> The base of the expression b^n is b and the exponent is n.

To develop the properties that allow us to simplify expressions with exponents, we will consider examples that allow you to use your calculator and then generalize the situation. Suppose we consider the following:

In the problem, $3^2 \cdot 3^7$, what does 3^2 represent? What does 3^7 represent?

$$3^2 \cdot 3^7 = (3 \cdot 3)(3 \cdot 3 \cdot 3 \cdot 3 \cdot 3 \cdot 3 \cdot 3)$$
$$= 3 \cdot 3 \cdot 3 \cdot 3 \cdot 3 \cdot 3 \cdot 3 \cdot 3 \cdot 3$$
$$= 3^9$$

We found $3^2 \cdot 3^7 = 3^9$. Notice that $2 + 7 = 9$.

Let's look at another example where the bases are the same variable. We multiply exponential expressions by adding the exponents. For instance, $x^2 \cdot x^7 = x^9$.

> We multiply exponential expressions containing the same base by adding the exponents.
>
> $$x^a \cdot x^b = x^{a+b}$$

We know that x^2 represents $x \cdot x$ and x^7 represents $x \cdot x \cdot x \cdot x \cdot x \cdot x \cdot x$. It follows that $x^2 \cdot x^7$ represents $x \cdot x \cdot x \cdot x \cdot x \cdot x \cdot x \cdot x \cdot x$, or x^9. This is $x^2 \cdot x^7 = x^{2+7} = x^9$.

Perform the indicated operations.

1. What is -3^2? $-3 \cdot 3 = -9$

2. What is $(-3)^2$? $(-3)(-3) = 9$

3. What is $(6x^4y^4)(4x^3 y^2)$?

$$
\begin{aligned}
(6x^4y^4)(4x^3y^2) &= (6 \cdot 4)(x^4 \cdot x^3)(y^4 \cdot y^2) \\
&= 24 \cdot x^{4+3} \cdot y^{4+2} \\
&= 24x^7y^6
\end{aligned}
$$

4. Multiply: $(-3a^4b^7)(8a^2 b^4)$

$$
\begin{aligned}
(-3a^4b^7)(8a^2b^4) &= (-3 \cdot 8)(a^4 \cdot a^2)(b^7 \cdot b^4) \\
&= -24 \cdot a^{4+2} \cdot b^{7+4} \\
&= -24a^6b^{11}
\end{aligned}
$$

5. Simplify $(x^4)(3xy)$.

$$
\begin{aligned}
(x^4)(3xy) &= 1x^4 \cdot 3xy \\
&= (1 \cdot 3)(x^4 \cdot x)y \\
&= 3 \cdot x^{4+1} \cdot y \\
&= 3x^5y
\end{aligned}
$$

If no coefficient is given as in x^4, it is understood that the variables have a coefficient of 1. If no exponent is given as in $3xy$, the exponents of both x and y are assumed to be 1. In both cases, we can write the 1 if needed with the understanding that the 1 does not change the value of the expression.

You Try It

Multiply the following:

1. $(-5)^2$

2. -5^2

3. $(4xy)(6xy)$

4. $(3ab)(-5ab)$

5. $(-7x^7y)(6x^3y^2)$

6. $(-x^2y^2z^2)(-3x^3z^4y^5)$

Division with Exponents

Suppose we examine $\dfrac{7^5}{7^3}$. What do 7^5 and 7^3 represent? We can calculate $\dfrac{7^5}{7^3}$ as follows:

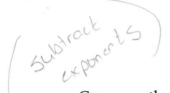

Subtract exponents

$$\frac{7^5}{7^3} = \frac{7 \cdot 7 \cdot \cancel{7} \cdot \cancel{7} \cdot 7}{\cancel{7} \cdot \cancel{7} \cdot \cancel{7}}$$
$$= 7 \cdot 7$$
$$= 7^2$$

Numbers are divided $\dfrac{49}{9a} \cdot \dfrac{5}{3} = 4a$

Compare the exponents of the original expression with the exponent of the result. You might notice that $5 - 3 = 2$. Also, $7^2 = 49$ and $\dfrac{7^5}{7^3} = \dfrac{16,807}{343} = 49$.

Next, let's try the same process with variables. For instance, $\dfrac{x^7}{x^3} = x^{7-3} = x^4$, and $\dfrac{x^7}{x^3} = \dfrac{x \cdot x \cdot x \cdot x \cdot \cancel{x} \cdot \cancel{x} \cdot \cancel{x}}{\cancel{x} \cdot \cancel{x} \cdot \cancel{x}}$. Since $\dfrac{x}{x} = 1$, we have $x \cdot x \cdot x \cdot x$, or x^4.

> We divide expressions with the same base by subtracting the exponent of the denominator from the exponent of the numerator.
>
> $$\frac{x^a}{x^b} = x^{a-b}, \text{ provided } x \neq 0$$

Complete the following:

1. What is the quotient of $\dfrac{6x^5}{2x^3}$?

$$\frac{6x^5}{2x^3} = \left(\frac{6}{2}\right)\left(\frac{x^5}{x^3}\right)$$
$$= 3x^{5-3}$$
$$= 3x^2$$

2. Simplify $\dfrac{6a^8b^7c^4}{2a^3b^4c^3}$.

$$\frac{6a^8b^7c^4}{2a^3b^4c^3} = \left(\frac{6}{2}\right)\left(\frac{a^8}{a^3}\right)\left(\frac{b^7}{b^4}\right)\left(\frac{c^4}{c^3}\right) = 3 \cdot a^{8-3} \cdot b^{7-4} \cdot c^{4-3} = 3a^5b^3c^1 = 3a^5b^3c$$

Remember $c^1 = c$.

$3a^5b^3c$

You Try It

Divide the following:

1. $\dfrac{40xy}{5xy}$

2. $\dfrac{-3a^3b^5}{15a^2b}$

3. $\dfrac{88a^{25}b^3}{-11a^{17}b}$

4. $\dfrac{-14x^5y^6z^7}{-30x^3z^4y^5}$

Reduce if larger # on the bottom

Zero Exponents

Simplify: $\dfrac{x^5}{x^5}$

Using the skills you learned in this unit, you might have answered x^0. Other students might have answered 1. This could create some serious discussions. Some problems result in zero exponents. However, in most math classes, answers with zero exponents are not considered to be in simplest form (except in scientific notation). So what should we do with zero exponents?

1.   Simplify each of the following using the skills you learned in this unit:

$$\frac{2^5}{2^5} = \quad , \quad \frac{(-3)^2}{(-3)^2} = \quad , \quad \frac{0.5^4}{0.5^4} = \quad , \quad \frac{100^3}{100^3} =$$

2.   Evaluate each of the original expressions using your calculator.

3.   Use your calculator to evaluate the following expressions:

$$2^0 = \quad , (-3)^0 = \quad , 0.5^0 = \quad , 100^0 =$$

4.   What happens when you try to evaluate 0^0? Why?

5.   Generalize the results by completing the equation below:

$$x^0 = \underline{\quad\quad}$$

Any nonzero number raised to the zero power is 1.

$$x^0 = 1, \text{ provided } x \neq 0$$

 Simplify as indicated.

1. Write $4x^3y^0z^5$ without zero exponents.

    $y^0 = 1$. Since $(4)(1) = 4$, the answer is $4x^3z^5$.

2. Simplify $(4a^3b^0c^5)(-2a^6b^4c^0d^0)$. The result should be written without any zero exponents.

    First, multiply the coefficients, then add the exponents as before to obtain $-8a^9b^4c^5d^0$. Since $d^0 = 1$, the simplified answer is $-8a^9b^4c^5$.

You Try It

Simplify the following. Express your answers without any zero exponents:

1. $(3x^2y^3)(2x^5y^3z^0)$

2. $(3r^5s^0)(2r^3t^5)$

3. $\dfrac{3m^5n^6t^2}{2m^0n^6p^0t}$

4. $\dfrac{3x^5y^5z^0}{2^0x^2y^5}$

Negative Exponents

What is $\dfrac{x^2}{x^7}$?

From before, we know that $\dfrac{x^2}{x^7} = x^{2-7} = x^{-5}$. But what does x^{-5} mean? Let's see if we can find a pattern that will help us understand negative exponents.

$$2^3 = \quad 8$$
$$2^2 = \quad 4$$
$$2^1 = \quad 2$$
$$2^0 = \quad 1$$
$$2^{-1} =$$
$$2^{-2} =$$
$$2^{-3} =$$

Looking at the results from each step, you might notice that $8 \div 2 = 4$. In each case to get from one step to the next, you divide by 2. So what is $1 \div 2$? In other words, $2^{-1} = \frac{1}{2}$. Similarly, $\frac{1}{2} \div 2 = \frac{1}{4} = \frac{1}{2^2}$. So $2^{-2} = \frac{1}{2^2}$.

Another way we can look at this situation is to examine the problem under consideration: $\frac{x^2}{x^7}$.

$$\frac{x^2}{x^7} = \frac{x \cdot x}{x \cdot x \cdot x \cdot x \cdot x \cdot x \cdot x} = \frac{1}{x \cdot x \cdot x \cdot x \cdot x} = \frac{1}{x^5}$$

Using what we learned earlier: $\frac{x^2}{x^7} = x^{2-7} = x^{-5}$

Thus, since we began with the same expression, $\frac{1}{x^5} = x^{-5}$.

> A base with a negative exponent can be written as one over the base with a positive exponent.
>
> $$x^{-n} = \frac{1}{x^n}, \text{ provided } x \neq 0$$

   Simplify each of the following. Write your answers without zero or negative exponents.

1.   What is $(7v^{-5})(-2v^3)$?

$$(7v^{-5})(-2v^3) = -14v^{-5+3} = -14v^{-2} = \frac{-14}{v^2}$$

2. What is $(5a^7b^{-5})(3a^{-4}b^{-3})$?

$$(5a^7b^{-5})(3a^{-4}b^{-3}) = 15a^{7+(-4)}b^{-5+(-3)} = 15a^3b^{-8} = \frac{15a^3}{b^8}$$

3. Simplify: $\dfrac{2x^3}{3x^{-3}}$

a) You could use the property learned earlier that $\dfrac{x^a}{x^b} = x^{a-b}$.

$$\frac{2x^3}{3x^{-3}} = \frac{2x^{3-(-3)}}{3} = \frac{2x^6}{3}$$

b) Alternately, this approach uses the property of exponents you just learned. Note that we inverted the divisor and multiplied the fractions.

$$\frac{2x^3}{3x^{-3}} = \frac{2x^3}{\frac{3}{x^3}} = 2x^3 \div \frac{3}{x^3} = 2x^3 \cdot \frac{x^3}{3} = \frac{2x^6}{3}$$

4. What is $(7v^5w^3)/(2v^{-3}w^3z^0)$?

$$\frac{7v^5w^3}{2v^{-3}w^3z^0} = \frac{7v^{5-(-3)}w^{3-3}}{2} = \frac{7v^8w^0}{2} = \frac{7v^8}{2}$$

5. Simplify: $\dfrac{-x^0}{x^{-4}}$

$\dfrac{-x^0}{x^{-4}} = -x^{0-(-4)} = -x^4$. Since $x^0 = 1$, we have $\dfrac{-x^0}{x^{-4}} = \dfrac{-1}{x^{-4}}$. Thus, $\dfrac{-1}{x^{-4}} = -x^4$.

A base with a negative exponent in the denominator can be written in the numerator as that base with a positive exponent.

$$\frac{1}{x^{-n}} = x^n$$

You Try It

Simplify the following. Express your answers with positive exponents only.

1. $(3x^{-2})(2y^3)$

2. $(3r^{-5})(-2r^3)$

3. $\dfrac{3t^{-2}}{2t^{-3}}$

4. $(-3x^{-2}y^{-5})(-2x^{-5}y^3)$

5. $\dfrac{3x^{-2}y^{-6}}{2x^{-5}y^8}$

6. $(-3a^0b^{-3}c^{-4})(4a^{-3}b^{-2}c^{-5})$

Zero Bases

You know that any number multiplied by zero is zero. Also, division by zero is undefined. You may have discovered that if the base is zero some of the exponent properties will not work. Why?

A zero base and positive exponent means repeated multiplication by 0.

1.   Simplify: 0^4 and 0^7

$$0^4 = 0 \cdot 0 \cdot 0 \cdot 0 = 0 \text{ and } 0^7 = 0 \cdot 0 \cdot 0 \cdot 0 \cdot 0 \cdot 0 \cdot 0 = 0$$

However, if zero is the base and the exponent is negative, then we may try to divide by zero.

2.   Simplify 0^{-4} and 0^{-7}.

$$0^{-4} = \frac{1}{0^4} = \frac{1}{0 \cdot 0 \cdot 0 \cdot 0} \text{ is undefined}$$

$$0^{-7} = \frac{1}{0^7} = \frac{1}{0 \cdot 0 \cdot 0 \cdot 0 \cdot 0 \cdot 0 \cdot 0} = \frac{1}{0} \text{ is undefined}$$

Properties of Exponents

Because of the difficulties with zero bases, we restrict the bases to avoid to division by zero. Notice the restrictions for the following properties. Discuss them with your group, and find an example that shows why each restriction is needed.

Properties of Exponents			
	Expression	**Simplification**	**Restriction**
1.	$x^a x^b$	x^{a+b}	
2.	$\dfrac{x^a}{x^b}$	x^{a-b}	$x \neq 0$
3.	$(x \cdot y)^a$	$x^a \cdot y^a$	
4.	$\left(\dfrac{x}{y}\right)^a$	$\dfrac{x^a}{y^a}$	$y \neq 0$
5.	$(x^a)^b$	x^{ab}	
6.	x^{-a}	$\dfrac{1}{x^a}$	$x \neq 0$
7.	x^0	1	$x \neq 0$
8.	$\dfrac{1}{x^{-a}}$	x^a	$x \neq 0$

Some of the properties in the table have not been covered in the text. Convince all members of your group that each is true using the same process that you did throughout the text.

You Try It

Simplify the following. Express your answers without negative or zero exponents.

1. $(3x^2 y^3)^3$

2. $(-3x^2 y^3)^{-2}$

3. $\left(\dfrac{3r^{-5} s^0}{t^{-5}}\right)^2$

4. $(3m^0 t^{-2})^2 (-2m^{-5} t^{-3})^2$

5. $\left(\dfrac{3x^{-2}}{y^{-5}}\right)^{-5}$

Multiplication of Variable Expressions

Look again at the geometric figures on the second page of this unit. Recall the following area formulas:

Rectangle: $A = L \cdot W$
Square: $A = s^2$
Triangle: $A = \frac{1}{2}bh$
Circle: $A = \pi r^2$

Find the area of each figure.

1. The area of this rectangle is $A = (2x)(3y) = 6xy$.

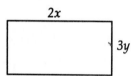

2. The area of this square is $A = (7z)^2 = 49z^2$.

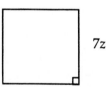

3. This quadrilateral is neither a rectangle nor a square. We have not discussed a formula for finding its area.

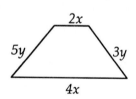

4. The area of this triangle is $A = \frac{1}{2}(2x)(5y) = 5xy$.

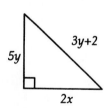

5. The area of the circle is $A = \pi(3a)^2 = \pi9a^2 = 9\pi a^2$.

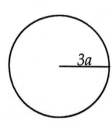

6.   Figure 6 is a rectangle. $A = (3x)(y + 5) = 3xy + 15x$.

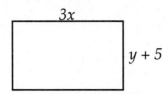

7.   The height is not given for the triangle so we cannot find the area.

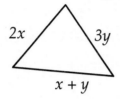

8.   The area of this triangle is $A = \frac{1}{2}(x + y)(2x - y)$.   We will simplify this expression later in this unit.

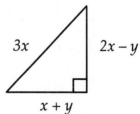

9. $A = (x + 3)(2x - y)$. We will simplify this expression for the area of this rectangle later in this unit.

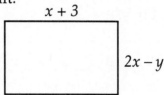

10.   Figure 10 is a pentagon. At this time, we do not have a formula to find its area.

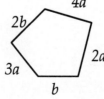

We will develop further methods for simplifying variable expressions similar to those in **8** and **9**.

More on the Distributive Law

We can use the distributive law to simplify expressions involving exponents.

> The Distributive Law: $a(b + c) = a \cdot b + a \cdot c$

 Simplify:

1. $3x(2x - 5)$

$$3x(2x - 5) = (3x)(2x) + (3x)(-5) = 6x^2 - 15x.$$

2. $3ab^2(2a^3b^{-5} - 5ab)$

$$3ab^2(2a^3b^{-5} - 5ab) = (3ab^2)(2a^3b^{-5}) + (3ab^2)(-5ab)$$
$$= 6a^4b^{-3} - 15a^2b^3$$
$$= \frac{6a^4}{b^3} - 15a^2b^3$$

You Try It

Multiply and simplify.

1. $3x^2y^3(2x^{-5}y^{-3} - 3xy)$ 2. $3r^{-5}s^0(2rt^{-5} + 5r^3t^{-4})$

3. $3m^0n^2t^{-2}(3m^0n^6t^{-2} - 2m^{-5}n^{-2}p^0t^{-3})$

Using the distributive law, we can now simplify the variable expressions from figures 8 and 9.

 1.   How would we multiply $(x + 3)(2x - y)$ to get the area of figure 9?

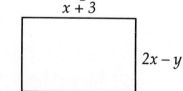

$$(x + 3)(2x - y) = (x + 3)(2x) + (x + 3)(-y)$$
$$= x \cdot 2x + 3 \cdot 2x + x(-y) + 3(-y)$$
$$= 2x^2 + 6x - xy - 3y$$

We distributed $x + 3$ over $2x$ and $-y$. Then, we used the distributive law to remove the two new parentheses around $x + 3$.

2. Multiply $\frac{1}{2}(x+y)(2x-y)$ to determine the area of figure 8.

$$\frac{1}{2}(x+y)(2x-y) = \frac{1}{2}[(x+y)(2x) + (x+y)(-y)]$$

$$= \frac{1}{2}[x \cdot 2x + y \cdot 2x + x(-y) + y(-y)]$$

$$= \frac{1}{2}(2x^2 + 2xy - xy - y^2)$$

$$= \frac{1}{2}(2x^2 + xy - y^2)$$

$$= x^2 + \frac{1}{2}xy - \frac{1}{2}y^2$$

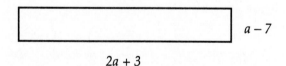

$3x$ $2x-y$

$x+y$

You Try It

1. Multiply using the distributive law: $(x + 5)(2x + 3)$

2. Multiply using the distributive law: $(x + 7)(x - 7)$

3. Find the area of the rectangle shown here.

$a - 7$

$2a + 3$

4. Find the area of a triangle with a base of $x + 5$ and a height of $2x + 3$.

The FOIL Method

To multiply two binomials, we use the distributive law. FOIL is a memory device that helps organize the multiplication of binomials. This insures that we multiply each term of the first binomial by each term of the second binomial. FOIL can only be used when we multiply one binomial by another.

The letters **FOIL** indicate that the product of two binomials is the sum of **F + O + I + L**. **F** represents the product of the first terms of the two binomials, and **O** represents the product of the outer terms. **I** represents the product of the inner terms of the binomials, and **L** represents the product of the last terms.

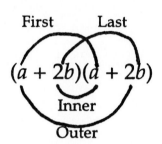

The product $(x + 5)(2x + 3)$ is the first problem from the last *You Try It*.

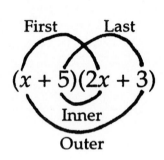

F   The first terms x and $2x$ have a product of $2x^2$.

O   The outer terms of x and 3 have a product of $3x$.

I   The inner terms of 5 and $2x$ have a product of $10x$.

L   The last terms of 5 and 3 have a product of 15.

$$F\ +O+\ \ I\ +\ L$$
$$\downarrow\quad\downarrow\quad\downarrow\quad\downarrow$$
$$(x+5)(2x+3)=2x^2+3x+10x+15=2x^2+13x+15$$

 Multiply $(3r - 3)(2r - s)$, another problem from the last **You Try It**.

F   The first terms ____ and ____ have a product of _____.

O   The outer terms of ____ and ____ have a product of _____.

I   The inner terms of ____ and ____ have a product of _____.

L   The last terms of ____ and ____ have a product of _____.

$$
\begin{array}{ccccccccc}
 & & & F & + & O & + & I & + & L \\
(3r-3)(2r-s) & = & & \underline{\hspace{1cm}} & + & \underline{\hspace{1cm}} & + & \underline{\hspace{1cm}} & + & \underline{\hspace{1cm}} \\
 & = & & \multicolumn{7}{l}{\underline{\hspace{5cm}}}
\end{array}
$$

You Try It

Problem	First Terms	Outer Terms	Inner Terms	Last Terms	Product
$(x+2)(2x-3)$	$(x)(2x)=$	$(x)(-3)=$	$(2)(2x)=$	$(2)(-3)=$	$2x^2-3x+4x-6=$
$(a-7)(2a+3)$					
$(v+3)(w+7)$					
$(x+2y)(x-2y)$					
$(3m+n)(2m-3n)$					

Multiplying Vertically

We can also multiply variable expressions vertically in a form similar to the one used for multiplication of whole numbers. This is especially useful when multiplying expressions with more than two terms.

Multiply $(x+y)(2x-y)$ vertically to find the area of this rectangle.

$$
\begin{array}{r}
x + y \\
2x - y \\
\hline
- xy - y^2 \\
2x^2 + 2xy \\
\hline
2x^2 + xy - y^2
\end{array}
$$

Here we multiplied x + y by −y.
Here we multiplied x + y by 2x
Combine the two rows of products.

$x+y$

$2x-y$

Now let's multiply a binomial by a trinomial. Since FOIL doesn't work, the vertical method is especially useful for these.

Multiply $(3x + 2y - 5)(2x - 4y)$.

$$
\begin{array}{r}
3x \ + \ 2y \ - \ \ 5 \\
2x \ - \ 4y \\
\hline
- \ 12xy \ - \ 8y^2 \ + \ 20y \\
6x^2 \ + \ 4xy \qquad\qquad\quad - \ 10x \\
\hline
6x^2 \ - \ 8xy \ - \ 8y^2 \ + \ 20y \ - \ 10x
\end{array}
$$

Here we multiplied $3x + 2y - 5$ by $-4y$.

Here we multiplied $3x + 2y - 5$ by $2x$.

Combine the two rows of the table.

Again we organized our work so that like terms were written in columns. This makes it easier to recognize and combine like terms.

Multiplying vertically is similar to multiplying whole numbers. Its advantages are that it helps organize your work and that it works for expressions with any number of terms.

You Try It

Multiply the following:

1. $(3t - 4)(2t - 3)$

2. $(a + b)(2a - 3b)$

3. $(x + 2y)(x - 2y)$

4. $(2x + 3)(3x - 5y + 2)$

5. $(a + b + 3)(2a - 3b - 2)$

Division of Variable Expressions

A monomial is a one-term variable expression. We can divide a variable expression by a monomial by separating the expression into fractions with each term of the numerator over the monomial.

 Simplify each of the following variable expressions.

1.   Simplify $\dfrac{x^2 + 5x}{x}$.

$$\frac{x^2 + 5x}{x} = \frac{x^2}{x} + \frac{5x}{x} = x + 5$$

The two terms of the numerator are used as the numerators of the separate fractions. Each fraction is reduced using the properties of exponents.

2.   Simplify $\dfrac{8w^3 + 4w^2}{4w^2}$.

$$\frac{8w^3 + 4w^2}{4w^2} = \frac{8w^3}{4w^2} + \frac{4w^2}{4w^2} = 2w + 1$$

3.   Simplify $\dfrac{7x^5 - 2x^3 + x^2}{x^2}$.

$$\frac{7x^5 - 2x^3 + x^2}{x^2} = \frac{7x^5}{x^2} - \frac{2x^3}{x^2} + \frac{x^2}{x^2} = 7x^3 - 2x + 1$$

4.   Simplify: $\dfrac{8x^3 - x^2 + 3x}{x^2}$

$$\frac{8x^3 - x^2 + 3x}{x^2} = \frac{8x^3}{x^2} - \frac{x^2}{x^2} + \frac{3x}{x^2}$$
$$= 8x - 1 + \frac{3}{x}$$

5.   If the area of a rectangle is $3x^2 - 9x$ and the length is $3x$, what is the width?

We need to determine the width of the rectangle so we first solve $A = L \cdot W$ for W, $W = \dfrac{A}{L}$. Substituting the given information and simplifying, we have $W = \dfrac{3x^2 - 9x}{3x} = \dfrac{3x^2}{3x} - \dfrac{9x}{3x} = x - 3$. So the width of the rectangle is $x - 3$.

You Try It

1. Divide and simplify: $\dfrac{4x^2 - x}{x}$

2. Divide and simplify: $\dfrac{25y^3 - 20y^2 + 10y}{10y}$

3. The area of the rectangle shown here is $3a^5 - 4a^3$. What is its length?

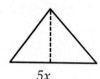

4. The height of a triangle is given by $h = \dfrac{2A}{b}$, where A is the area and b is the base. The area of the triangle is $5x^2 - 15x$. If the base is $5x$, what is the height?

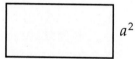

Scientific Notation

Sometimes we must work with very large or very small numbers. We usually express these numbers in scientific notation. Here are some examples:

One Billion Dollars	1.0×10^9 Dollars
One Year	3.2×10^7 Seconds
Distance From Earth to the Moon	3.8×10^8 Meters
Mass of Mars	6.24×10^{23} Kilograms
Nanosecond	1.0×10^{-9} Seconds
Power required for a T1 Line	3.0×10^{-6} Watts
Speed of TV Signals	3.0×10^8 Meters per Second
Atomic Mass Unit	1.66×10^{-27} Kilogram

Each of these numbers is expressed as a number greater than or equal to 1 and less than 10 multiplied by a power of ten. This is called scientific notation. We do not simplify negative and zero exponents when we work in scientific notation.

> A number written in **scientific notation** has the form
> $$M \times 10^E,$$
> where $1 \leq M < 10$ and E is an integer.

The numbers in the next table are expressed in both decimal and scientific notation. There is a connection between the position of the decimal point of the number in decimal notation and the power of ten in scientific notation. What is it?

Decimal Notation	Scientific Notation
3,000,000	3.0×10^6
0.00085	8.5×10^{-4}
750,000	7.5×10^5
0.00000000000044	4.4×10^{-13}
3.889	3.889×10^0
0.9	9.0×10^{-1}
90	9.0×10^1

Look at 3,000,000 and 3.0×10^6. How many places and in what direction would you move the decimal to change 3.0 to 3,000,000? _____ How many times would we multiply 3.0 by 10 to get 3,000,000? _____ Compare your answers.

Look at 0.00085 and 8.5×10^{-4}. How many places and in what direction would you move the decimal to change 8.5 to 0.00085? _____ How many times would we divide 8.5 by 10 to get 0.00085? _____ Compare your results.

You Try It

Complete the following table.

	Decimal Notation	Scientific Notation
1.	750,000	
2.	0.00407	
3.		3.11×10^0
4.		2.03×10^6
5.		1.04×10^{-8}
6.		1.25×10^{11}
7.	0.45	
8.	4.5	
9.	45	

Operations in Scientific Notation

We use properties of exponents to multiply or divide numbers expressed in scientific notation. To multiply, we add the exponents. When we divide, we subtract the exponents.

 Simplify the following:

1. $(3.5 \times 10^5)(2.0 \times 10^4) = 7 \times 10^9$

   Multiply the numbers and add the exponents.

   $$(3.5)(2.0) = 7, \text{ and } (10^5)(10^4) = 10^{5+4} = 10^9$$

2. $(5.5 \times 10^{-3})(4.4 \times 10^{-7}) = 24.2 \times 10^{-10} = 2.42 \times 10^{-9}$

   $$24.2 = 2.42 \times 10^1, \text{ and } (10^1)(10^{-10}) = 10^{-9}$$

3. $\dfrac{8.8 \times 10^{22}}{4.4 \times 10^{17}} = 2 \times 10^5$

   Divide the numbers and subtract the exponents.

   $$\frac{8.8}{4.4} = 2, \text{ and } \frac{10^{22}}{10^{17}} = 10^{22-17} = 10^5.$$

4. $\dfrac{9.9 \times 10^{22}}{3.3 \times 10^{-7}} = 3 \times 10^{29}$

   $$\frac{9.9}{3.3} = 3, \text{ and } \frac{10^{22}}{10^{-7}} = 10^{22-(-7)} = 10^{29}.$$

You Try It

1. $(6.6 \times 10^{-7})(2.2 \times 10^4) =$

2. $(4.4 \times 10^{-3})(2.5 \times 10^{-7}) =$

3. $(1.1 \times 10^{77})(2.8 \times 10^{34}) =$

4. $\dfrac{7.07 \times 10^{-12}}{1.01 \times 10^{-7}} =$

5. $\dfrac{6.66 \times 10^{12}}{3.33 \times 10^{-7}} =$

Estimation

If you first estimate the answer before using a calculator, you can compare your estimate with the calculator result to determine if it is reasonable. When estimating, change the numbers to scientific notation and then round to one digit. One digit allows for mental multiplication and is sufficient for most cases.

1. A rectangular field is 3754 meters wide by 7893 meters long. Estimate its area.

    Round: $3754 \approx 4 \times 10^3$ and $7893 \approx 8 \times 10^3$

    Estimate: $(4 \times 10^3)(8 \times 10^3) = 32 \times 10^6 = 3.2 \times 10^7$

    Why did we change 32 to 3.2? Did the exponent change? The area is approximately 3.2×10^7 square meters. Write this area in decimal notation.

2. An office building with 385,000 square feet of floor space sold for $33,000,000. Estimate the price per square foot.

    Round: $385,000 \approx 4 \times 10^5$ and $33,000,000 \approx 3 \times 10^7$

    Estimate: $\dfrac{3 \times 10^7}{4 \times 10^5} = 0.75 \times 10^2 \text{ \$/ft}^2 = 7.5 \times 10^1 \text{ \$/ft}^2$

    Why did the 0.75 change to 7.5? Did the exponents change? The office building sold for approximately $7.5 \times 10^1 = 75$ dollars per square foot.

You Try It

Estimate answers to the following:

1. Each year the United States shreds 1.4×10^7 pounds of worn-out paper currency. If each bill is worth one dollar and 453 dollars weigh 1 pound, how much money is destroyed?

2. $\dfrac{356,078,000}{0.00009876}$

3. $(257,000)^4$

Scientific Notation on Your Calculator

In science or technology classes, you may discuss significant digits and rounding. In this course, we will round the final answer to two decimal places in scientific notation.

Evaluate $(6.67 \times 10^{-7})(2.72 \times 10^4)$ on your calculator.

First, estimate the answer.

$$(6.67 \times 10^{-7})(2.72 \times 10^4) \approx (7 \times 10^{-7})(3 \times 10^4)$$
$$= 21 \times 10^{-3}$$
$$= 2.1 \times 10^{-2}$$

Round to one digit

Calculate estimate

We should expect the calculator answer to be close to 2.1×10^{-2}.

To work the problem on your calculator:

Press MODE, arrow over to **Sci**, and press ENTER to put your calculator in scientific notation mode. Now press 2nd MODE [**QUIT**] to return to the home screen.

Press (6 . 6 7 2nd , [**EE**]. The screen should show (6.67 ε at this point. Press (-)7) to complete the first factor. Press (2.72 [**EE**] 4) to enter the second factor in scientific notation. Your screen should look like this one.

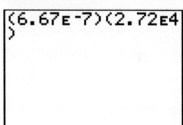

Press ENTER. The answer appears as 1.81424 ε -2, which is the calculator equivalent to 1.81424×10^{-2}. The calculator uses ε to represent the base 10 for numbers expressed in scientific notation. This is 1.81×10^{-2} when rounded to two decimal places.

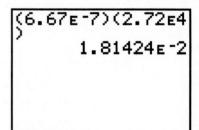

Compare your estimate with your calculator answer. Is the calculator answer of 1.81×10^{-2} close enough to the estimate 2.1×10^{-2}?

Use your calculator to evaluate: $\dfrac{6.67 \times 10^{-7}}{2.72 \times 10^4}$

First, estimate the answer.

$$\dfrac{6.67 \times 10^{-7}}{2.72 \times 10^4} \approx \dfrac{7 \times 10^{-7}}{3 \times 10^4}$$

$$= \dfrac{7 \times 10^{-7}}{3 \times 10^4} \qquad\qquad \textit{Round to one digit}$$

$$= \dfrac{7}{3} \times 10^{-11} \qquad\qquad \textit{Calculate estimate}$$

$$\approx 2.3 \times 10^{-11} \qquad\qquad \textit{Change to decimal}$$

We expect the calculator answer to be close to 2.3×10^{-11}.

Now work the problem on your calculator.

Press ⌶⌶6⌶.⌶6⌶7⌶ ⌶2nd⌶⌶,⌶[EE]. The screen should show (6.67 E at this point. Press ⌶(-)⌶7⌶)⌶. Press ⌶÷⌶. Press (2.72 [EE] 4) to enter the divisor of the calculation. Your screen should look like this one.

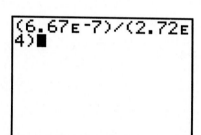

Press ⌶ENTER⌶. On the calculator, the answer appears as 2.45220588 E⁻11, which is 2.45×10^{-11} rounded to two decimal places.

 Compare the estimate and the calculator answer. Is the calculator answer 2.45×10^{-11} close enough to the estimate 2.3×10^{-11}?

What did you or your group decide about the "closeness" of the estimates and answers for the last two problems? Did the estimate show that the calculator answer was reasonable? There is no rule to determine closeness. However, you should check to see if the exponents and the numbers are close.

You Try It

1. A rectangular field is 3754 meters wide by 7893 meters long. Earlier we estimated the area to be 32,000,000 sq. meters. Use your calculator to find the area, and compare your answer with the estimate.

2. An office building with 385,000 square feet of floor space sold for $33,000,000. Earlier in this unit, we estimated the cost to be $75 per square foot. Find the actual cost and compare your answer with the estimate.

Evaluate the following using scientific notation on your calculator.

$$\frac{(1.67 \times 10^8)^4}{(3.72 \times 10^{-5})^3} =$$

First, estimate the answer.

$$\frac{(1.67 \times 10^8)^4}{(3.72 \times 10^{-5})^3} \approx \frac{(2 \times 10^8)^4}{(4 \times 10^{-5})^3} \qquad \textit{Rounding}$$

$$= \frac{(16 \times 10^{32})}{(64 \times 10^{-15})} \qquad \textit{First Estimate}$$

$$= 0.25 \times 10^{47} \qquad \textit{Simplifying}$$

$$= 2.5 \times 10^{46} \qquad \textit{Change to scientific notation}$$

We expect the calculator answer to be close to 2.5×10^{46}.

Now work the problem on your calculator.

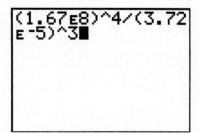

Enter the (1.67 ε 8) as before. Press $\boxed{\wedge}\boxed{4}$ to raise everything in the parentheses to the 4th power. Press $\boxed{\div}$ followed by (3.72 ε ⁻5). Use the $\boxed{\wedge}$ key to cube the expression.

Press ENTER. The answer is 1.51×10^{46} rounded to two decimal places. Your screen should look like the one to the right.

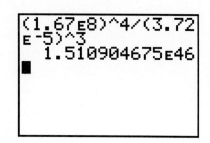

 Compare the estimate with your calculator answer. Is the calculator answer of 1.51×10^{46} close enough to the estimate 2.5×10^{46}?

You Try It

Estimate the answer, and then use your calculator to work the problem. Place a ✓ in the last column if your estimate shows that the calculator answer is reasonable. Write calculator answers in scientific notation rounded to two decimal places:

	Problem	Estimate	Calculator Answer	Check
1.	$(9.67 \times 10^{-17})(3.72 \times 10^{44})$			
2.	$(7.67 \times 10^{-7})^3$			
3.	$\dfrac{(1.67 \times 10^{-7})(7.67 \times 10^{-7})}{(9.99 \times 10^{-15})}$			
4.	$(8.939 \times 10^{-22})^{-2}$			

Summary

During your study of this unit, you have:

1. Calculated areas and perimeters of geometric figures with dimensions given as variable expressions;

2. Simplified variable expressions by combining like terms;

3. Simplified variable expressions with integer exponents using the properties of exponents;

Properties of Exponents

Expression	Simplification	Restriction
$x^a x^b$	x^{a+b}	
$\dfrac{x^a}{x^b}$	x^{a-b}	$x \neq 0$
$(x \cdot y)^a$	$x^a \cdot y^a$	
$\left(\dfrac{x}{y}\right)^a$	$\dfrac{x^a}{y^a}$	$y \neq 0$
$(x^a)^b$	x^{ab}	
x^{-a}	$\dfrac{1}{x^a}$	$x \neq 0$
x^0	1	$x \neq 0$
$\dfrac{1}{x^{-a}}$	x^a	$x \neq 0$

4. Used the properties of exponents and the distributive law to multiply variable expressions;

5. Multiplied using the distributive law;

6. Divided variable expressions with monomial denominators;

7. Used scientific notation for calculations involving very large or very small numbers;

8. Compared estimates with calculator answers;

9. Used the calculator for calculations not easily done by pencil and paper; and

10. Used the following TI-83 features:

MODE to change to scientific notation mode.
2nd , [**EE**] to enter numbers in scientific notation.

Unit 3 Problems for Practice

A. Solve the following equations algebraically and graphically on the calculator.

1. $3x - 8 = 15$

2. $7 - 3c = -17$

3. $5y - 11 = 3y + 13$

4. $-5 + 11d = 16 - 5d$

5. $2(3x - 5) = x + 7$

6. $-3(2b - 11) = 6 - 2b$

7. $2(3c + 5) + 7(2c - 1) = 3(4 - 8c)$

8. $5 - (3k + 44) = -6 + 11(5k - 7)$

9. The Texas chili recipe calls for $1\frac{1}{2}$ tablespoons of chili powder C for each pound of meat m. Marie adds one more for the pot for just the right spiciness, $C = \frac{3}{2}m + 1$. If Marie has only 7 tablespoons of chili powder, how many pounds of meat does she need to use?

10. After bowling two games, Shadrach's average score was 107 points. What must Shadrach bowl in his third game to average 110 points for the three games?

B. Solve the following equations algebraically and numerically on the calculator.

11. $2x - 8 = 15$

12. $7 - 3c = -1$

13. $4y - 11 = 7y + 13$

14. $5 - 11d = 16 - 5d$

15. $-2(3x - 5) = x - 7$

16. $3(5b - 11) = 6 - 2b$

17. $3(3c + 5) - 7(2c - 1) = 3(4 - 8c)$

18. $-11 - (3k - 44) = -6 + 11(5k - 7)$

19. In 1982, the Department of Transportation ordered that the maximum speedometer reading be 85 mph. If the number of miles per hour M and the number of kilometers per hour K are related by the equation $M = 0.621K$, what is the maximum allowed speed in kilometers per hour?

20. The printing costs for the cookbook of a local organization are $3.50 per book. The selling price is $14.95. Operational costs last year for the organization were $2250. The profit P the organization makes per book b is $P = 14.95b - 3.50b$. How many books must the organization sell to cover their operational costs?

C. Interpret each graph to answer the questions.

21. The following graph shows the stopping distance as a function of the speed of the car. Complete the input-output table below.

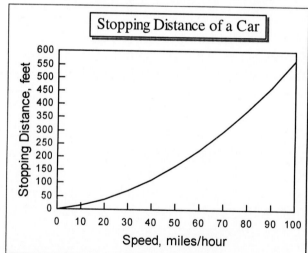

Speed	Stopping Distance
40 mph	
55 mph	
60 mph	
	375 ft
	75 ft

a) What is the stopping distance for the input that is the legal speed limit on a residential street in your town?

b) What is the stopping distance that is the output if the input is the legal speed limit on an interstate in your state?

22. This graph shows college tuition as a function of the academic year for public universities and colleges for in-state students. Complete the input-output table below.

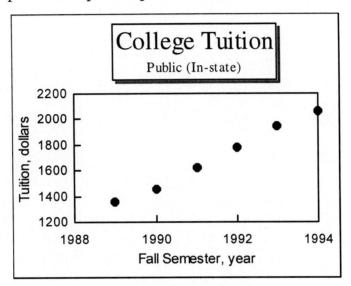

Year	Dollars
1989	
1994	
	$1782
	$1624

a) What was the increase in tuition between 1990 and 1992?

b) What was the increase in tuition between 1992 and 1994?

c) Was the increase in tuition larger between 1990 and 1992 or between 1992 and 1994? Discuss how you might use the graph to answer this question.

D.   Make I-O tables for the following and graph on graph paper.

23. $y = 2x - 3$ 24. $y = 3x + 2$

25. $y = \frac{1}{2}x - 4$ 26. $y = \frac{-2}{3}x + 3$

E.   Create at least four equations for each graph in part D and solve using the graphs you created.

F.   Make I-O tables for each of the following:

27.   In 1994, the Leaning Tower of Pisa was about 5.18 meters off perpendicular with the lean increasing by 1.25 millimeters a year. The amount of lean L can be calculated by the equation $L = 5.18 + 0.00125t$, where t is the number of years after 1994. How much will the tower be leaning for each of the years from 1994 until 2000?

28.   After 1990, the number of active physicians P in the U.S. per 100,000 people is given by the equation $P = 2.44t - 4632$, where t is the year. How many active physicians are predicted for each of the years from 1996 until 2001?

29.   Complete an I-O table for the relationship between a man's weight W (in pounds) and his height h (in inches) given by the equation $W = 5h - 190$. Complete the table for men who are between 5 feet 6 inches and 6 feet 2 inches tall.

30. A growing child age *A* needs *H* hours of sleep according to the equation $H = 13 - 0.29A$ rounded to the nearest hour. Complete an I-O table for ages 7–12. What time should an 8-year-old go to bed to get up at seven in the morning to go to school?

G. Find the perimeter of the following geometric figures.

31.

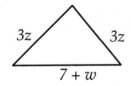

3y

2x   2x

3y

32.

5l

3w   3w

5l

33.

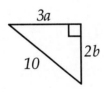

3z   3z

7 + w

34.

x   1.5y

2y

   y

4x

35.

3a

10   2b

36.

1.7b

37.

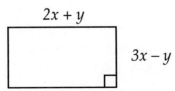

2x + y

3x − y

38.

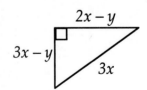

2x − y

3x − y

3x

3148

H.   Simplify the following by combining like terms.

39. $3x + 5 - 4 + 8x$

40. $5 - 2y - 11 - 3y$

41. $2t + 3s - 5t - 11s$

42. $2a + 3b - 11 - 5a - 7b - 5$

43. $6x^2 + 3x - 4x^2 + 1$

44. $5x - 7 - x^3 - 4x + 3$

45. $7c^3 - 5c + 2 - 4c^3 + 8c - 7$

46. $5 - d - d^4 + 11 - 5d - 3d^4$

47. $4(2x + 3) - 5(7 - 2x)$

48. $-3(2w - r) + 4(5r - 3w)$

49. $4(2x + 3y) - 5(2x - 3) - 2y$

50. $3(8j - 4) - 5(-2k - 4) - 3(2j + 2k)$

I. Find the area of the following geometric figures.

51.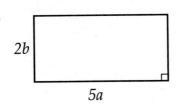

52.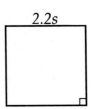

53.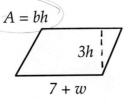

54.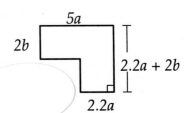

55.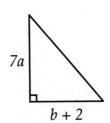

56.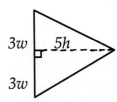

57.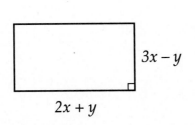

58.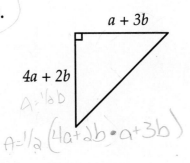

J. Expand and simplify, if possible:

59. $3x(2x+3)+2x(2x+3)$

60. $-2a(4-3a)-5(4-3a)$

61. $2c(3c+d)-2d(3c+d)$

62. $-5k(2n+3r)+4r(2n+3r)$

63. $(2x-3)(3x-7)$

64. $(2a-4b)(3a+2b)$

65. $(5-3v)(5+3v)$

66. $(9a-11b)(9a+11b)$

67. $(2x-7)(3x-4y+8)$

68. $(3b-4c)(2b-4c+11)$

K. Write the following without negative or zero exponents:

69. $(3x^2)(x^3)$

70. $(3a^2)(5b^3)$

71. $(3x^2y^2)(-5x^5y^3)$

72. $(-3c^5d^2)(-5c^3d^3)$

73. $(-2a^3b^4)(6a^4b)$

74. $(4cd)(12c^5d^7)$

75. $(-7r^4t^6z)(-5r^6tz^5)$

76. $(x^7y^{11}z^{21})(x^{13}y^{19}z^{29})$

77. $(3a^2)(5b^{-3})$

78. $(-4r^{-2}t^{-2})(8r^5t^{-3})$

79. $x^{-1}(x^2+y)$

80. $x(x^{-2}+2y)$

81. $(3x)^{-2}$

82. $3x^{-2}$

83. $(-4x)^0$

84. $-4x^0$

85. $(-4a^3)^{-3}$

86. $\left(\dfrac{-4}{a^3}\right)^{-4}$

87. $5x^2y(z^0x^{-2}y^4+1)$

88. $-3z^{-3}(px^2-5)$

89. $9ab(1.2a^{-7}-1)$

90. $ab^2c^{-3}(4a^{-1}v^{-2}c^3)$

91. $pq^{-3}(p+q+pq^{-3})$

92. $-3m^2n^{-2}(-2m^{-1}+n^2+5n^{-4})$

93. $x^{-10}y^7z^{22}(x^{11}y^{-8}z^{-22}+11xyz)$

94. $-3x^0$

95. $-5x^3$

96. $(-3x)^0$

L. Divide and simplify:

97. $\dfrac{x^2 - 4x}{x}$ **98.** $\dfrac{40y^4 - 20y^3 + 10y^2}{10y^2}$

99. $\dfrac{3a^5 - 4c^3}{a^2}$ **100.** $\dfrac{99r^8 - 66r^4 + 33r^2}{11r^2}$

101. $\dfrac{w^3 + 5w^2z}{wz}$ **102.** $\dfrac{7x^5 - 2x^3 + x^2}{x^2}$

103. $\dfrac{25p^4 + 10p^3 - 5p^2}{5p^2}$ **104.** $\dfrac{2y^3 - 6y^6 + 18y^9}{2y}$

105. $\dfrac{8x^7 - 12x^5 + 6x^2}{2x^3}$ **106.** $\dfrac{800i^7 - 120i^5 + 4000i^4}{40i^4}$

M. Change each of the following to scientific notation.

107. 0.000935 **108.** 345

109. 664,200 **110.** 0.00345

N.   Change each of the following to decimal notation.

111. 2.35×10^{-4} **112.** 2.78×10^{0}

113. 5.78×10^{6} **114.** 7.23×10^{-3}

O.   Work the following without using your calculator. Your final answer should be in scientific notation.

115. $(210,000)(77,000,000)$ **116.** $(0.000007)(0.00000011)$

117. $\dfrac{77,000,000}{0.00000011}$ **118.** $\dfrac{(210,000)(77,000,000)}{0.00000011}$

119. $(22,000,000,000)^{2}$

P.   For each of the following:
    a)   Write the original exercise in scientific notation.
    b)   Use scientific notation to estimate the answer.

120. $(230,000,000)(679,000,000,000)$

121. $(0.0000000777)(0.0000000333)$

122. $\dfrac{(140,003)(259,001,006)}{(0.000093)}$

123. $(3,020,000,000,000)^{2}$

124. $\dfrac{(210,000)(77,000,000)}{(0.00000011)(0.000005)}$

125. $(40,700,300,000)^{-3}$

Q.   For each of the following:
   a)   Estimate the answer;
   b)   Evaluate using your calculator. Give answers in scientific notation, correct to two decimal places;
   c)   Compare answers.

126. $(217,000)(737,000,000)$

127. $(0.0000307)(0.000001011)$

128. $\dfrac{0.000001011}{727,000,000}$

129. $(3,020,400,000,000)^3$

130. $\dfrac{(5,210,000)(727,000,000)}{0.000003011}$

131. $(0.00050022)^3(0.0000700044)^2$

132. $\dfrac{(0.000003011)}{(210,005,000)(4,077,000,000)}$

133. $\dfrac{(210,005,000)(4,077,000,000)}{(0.000003011)(0.0000091005)}$

R.   Application problems using scientific notation

134.   The radius of the sun is 695,990 kilometers, or about 700,000 kilometers. Write each of these in scientific notation.

135.   The nucleus of a human liver cell is approximately 0.000004 meters in diameter. Write this in scientific notation.

136.   An amoeba is about 0.00008 meters across. To do a line dance, 68,000,000 amoebas are placed side-by-side. How long is the line?

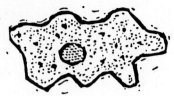

137.   The moon orbits the earth approximately 12 times a year. The length of the orbit of the moon around the earth is approximately 2,400,000 kilometers. How far will the moon travel in a year?

138.   The smallest living cells, called *Mycoplasma*, are 0.0000002 meters in diameter. If 39,000,000,000 of these cells are placed on top of each other, how high is the pile of cells?

139.   A standard 3.5" high density computer diskette can hold 1,400,000 bytes of information. How many bytes are on 5500 diskettes?

140.   Some scientists think the universe is about 15,000,000,000 years old. How many seconds would this be if expressed in scientific notation?

141.   The plan to repair the stonework in York Minster Cathedral in England will cost three million pounds, or $4,500,000. Express the ratio of the cost in pounds to the cost in dollars in scientific notation.

142.   A convention center has a large display room that is 1745 feet long by 2555 feet wide. A new industrial grade carpet will cost $9.75 per square yard for the carpet plus $2.10 per square yard for installation. What is the cost of carpeting the display room?

143. In the July 1991 *Natural History* magazine, Lewis T. Nielsen reported, "In one experiment, Canadian researchers who uncovered their torsos, arms and legs to arctic mosquitoes in the interest of science, reported as many as 9000 bites per minute. At this rate, unprotected humans would lose half of their blood supply in two hours, easily enough blood loss to cause death."

    a) In scientific notation, how many times would an unprotected human be bitten in one hour?

    b) If the approximate volume of blood removed per bite is 2.3 μL (microliters) or 2.3×10^{-6} L, how much blood loss could be expected in one minute? Report your answer in scientific notation.

    c) Compare the amount of blood loss in 2 hours with the average human blood volume of 5.5 L.

144. One gallon of carelessly discarded motor oil can contaminate 1 million gallons of groundwater. This is a one-year supply for 50 people.

    a) In scientific notation, how many gallons of water does each person need for one year?

    b) In 1993, there were approximately 1.9×10^8 motor vehicles registered in the U.S. If each vehicle was responsible for a spill of 0.1 gallons of oil a year, what was the potential water contamination?

    c) The world population is predicted to be 6.3×10^9 by the year 2000. How much water will be needed to support this many people?

145. Look up the number of people in the United States and the federal debt for the same year. What is the average federal debt per person in the United States for that year?

146.   The *Pathfinder* landed on Mars on July 4, 1997. According to one news report, it traveled through space at 47,500 miles per hour and the *Pathfinder* flew 4.97×10^8 miles as it arced its way around the solar system to Mars.

a)   If this information is correct, how many days did the journey take?

b)   How many days did it really take? Assuming the speed was correct, how far did the *Pathfinder* travel?

Linear Equations and Inequalities

Upon successful completion of this unit you should be able to:

1. Solve fractional equations;

2. Solve proportions;

3. Solve direct variation equations;

4. Solve percent equations;

5. Solve literal equations;

6. Solve inequalities in one variable;

7. Write inequalities in both interval and inequality notation;

8. Graph inequalities on a number line; and

9. Solve application problems.

Introduction

In Unit 2, we solved linear equations numerically, graphically, and algebraically. We can use linear equations to solve a wide variety of problems.

For example, fractional equations can be used to solve circuit and rate problems. Proportions can be used to solve map and medical dosage problems. Percent equations can be used to solve problems involving markup and markdown. Direct variation equations can be used to solve problems involving distance and sales tax. You will see other examples in the unit and problem set.

We will also solve inequalities. These arise when we need to answer questions involving *less than*, *more than*, *at least*, or *at most* rather than equals. We will learn to express solutions in inequality notation, interval notation, and on number line graphs.

Let's begin with an equation with fractions.

Fractional Equations

To reduce the number of possible errors, we often solve an equation with fractions by multiplying both sides of it by the least common multiple (LCM) of the denominators. Multiplying by the LCM eliminates the fractions and usually simplifies the equation to be solved. Multiplying both sides of the equation by the same number does not change the answer.

One application that uses fractional equations is resistance in a circuit.

Resistance in a Circuit

Suppose a circuit has two resistors in parallel with a total resistance of 3 ohms. One of the resistors has a value of 4 ohms. What is the value of the other resistor?

The total resistance of this circuit is written as:

$$\frac{1}{R} = \frac{1}{R_1} + \frac{1}{R_2}$$

where R is the total resistance, R_1 is the value of the first resistor, and R_2 is the value of the second resistor.

Substitute 3 for the total resistance R and 4 for the first resistor R_1:

$$\frac{1}{3} = \frac{1}{4} + \frac{1}{R_2}$$

This fractional equation has denominators of 3, 4, and R_2. The least common multiple of the denominators is $12R_2$.

Multiply both sides of the equation by $12R_2$ to eliminate the fractions, then solve for R_2.

$$12R_2\left(\frac{1}{3}\right) = 12R_2\left(\frac{1}{4} + \frac{1}{R_2}\right)$$

$$12R_2\left(\frac{1}{3}\right) = 12R_2\left(\frac{1}{4}\right) + 12R_2\left(\frac{1}{R_2}\right)$$

$$4R_2 = 3R_2 + 12$$

$$R_2 = 12$$

The value of the second resistor is 12 ohms.

A Rate Problem

Two streams are emptying into a reservoir after a sudden rain storm. The flow is such that the reservoir will be full after two hours. If the flow from the first stream alone would fill the reservoir in three hours, how many hours would it take the second stream alone to fill the reservoir?

How much of the reservoir could the first stream fill in one hour? The rate of the first stream is $\frac{1}{3}$ reservoir per hour. Why?

Let t represent the time required for the second stream to fill the reservoir. The rate of the second stream is $\frac{1}{t}$ reservoir per hour. Why?

Both streams flow into and fill the reservoir after two hours. We calculate each stream's flow by multiplying the rate of the stream by 2 hours. At the end of 2 hours, we will have 1 reservoir full of water so we set the sum of the flows equal to 1. The following equation gives their combined flow:

$$2\left(\frac{1}{3}\right) + 2\left(\frac{1}{t}\right) = 1$$ *Multiply the time by the rate to find the flow.*

$$\frac{2}{3} + \frac{2}{t} = 1$$ *Simplify.*

The LCM of 3 and t is _____.

$$3t\left(\frac{2}{3} + \frac{2}{t}\right) = 3t(1)$$ *Multiply by LCM.*

$$2t + 6 = 3t$$
$$6 = 3t - 2t$$
$$6 = t$$

The second stream alone would fill the reservoir in 6 hours.

 Solve the flow equation numerically and/or graphically on your calculator. Does your calculator answer agree with the one above?

You Try It

Solve the following fractional equations. First, multiply both sides by the LCM of the denominators.

1. $\dfrac{2}{3} + \dfrac{x}{4} = 5$

2. $\dfrac{1}{4} = \dfrac{1}{r} + \dfrac{1}{5}$

3. $\dfrac{7}{12} + \dfrac{x}{3} = \dfrac{5}{4} - \dfrac{x}{2}$

4. Find the value of the second resistor in a parallel circuit with total resistance of 2 ohms and one resistor of 3 ohms.

5. Two streams are emptying into a reservoir after a sudden rain storm. The flow is such that the reservoir will be full after two hours. If the flow from the first stream alone would fill the reservoir in five hours, how many hours would it take the second stream alone to fill the reservoir?

Ratios and Proportions

Equations with one ratio on each side are a special type of fractional equation called **proportions**. We can solve the next two problems using proportions.

Scale on a Map

The legend on a map states that the scale is 1 inch to 50 miles. You measure the distance between two cities and find that they are 3.5 inches apart on the map. How far apart are the cities?

Setup the ratios for the scale:　　1 inch = 50 miles ➔ $\dfrac{1 \text{ inch}}{50 \text{ miles}}$

Setup the ratio for the unknown: 3.5 inches = m miles ➔ $\dfrac{3.5 \text{ inches}}{m \text{ miles}}$

The ratios are equal. This leads to the proportion:

$$\frac{1 \text{ inch}}{50 \text{ miles}} = \frac{3.5 \text{ inches}}{m \text{ miles}}$$

$$50m\left(\frac{1}{50}\right) = 50m\left(\frac{3.5}{m}\right) \qquad \textit{Multiply by LCM.}$$

$$m = 175 \text{ miles}$$

The cities are 175 miles apart.

While the method we used to solve fractional equations can be used to solve proportions, they can also easily be solved using cross multiplication. To cross multiply, we multiply the first numerator by the second denominator and the first denominator by the second numerator. The two products are set equal to each other and solved.

$$1 \cdot m = (50)(3.5)$$
$$m = 175 \text{ miles}$$

The cities are 175 miles apart.

> You can solve proportions
> by cross multiplying:
>
> If $\dfrac{a}{b} = \dfrac{c}{d}$, then $ad = bc$

The method of cross multiplication is usually faster; however, it works only for proportions, that is, equations with exactly one fraction on each side.

A Dose of Medicine

The dosage of a certain medication is determined by the patient's body weight. Each fifty pounds of body weight requires 3.5 milliliters of the medicine. How much should a 175-pound person take?

Let n be the number of milliliters the 175-pound person is to take.

Setup the ratio for the dosage: 3.5 ml for 50 lb.　➜　$\dfrac{3.5 \text{ ml}}{50 \text{ lb.}}$

Setup the ratio for the unknown: n ml for 175 lb.　➜　$\dfrac{n \text{ ml}}{175 \text{ lb.}}$

Set the ratios equal to each other, then solve by cross multiplying.

$$\frac{3.5 \text{ ml}}{50 \text{ lb.}} = \frac{n \text{ ml}}{175 \text{ lb.}}$$

$$(3.5)(175) = 50n$$

$$n = 12.25 \text{ ml}$$

A 175-pound person would take 12.25 milliliters of the medicine.

You Try It

Solve the following:

1. $\dfrac{14}{x} = \dfrac{22}{35}$

2. $\dfrac{(x+5)}{4} = \dfrac{3}{7}$

3. Two cities are 185 miles apart. The legend on a map indicates that the scale is 1 inch to 50 miles. How far apart would they be on the map?

4. Each fifty pounds of body weight requires 3.5 milliliters of the medicine. How much of the medicine should a 250-pound person take?

5. One mile is equal to 5280 feet. How many feet are in 3.5 miles? How many miles are in 17,450 feet?

For these problems, a scale or a ratio was given that allowed us to write a proportion because we knew the relationship between the variables. Next, we'll look at applications of direct variation that have a different type of relationship.

Direct Variation

Direct variation problems can lead to linear equations. If we know that x varies directly as y, we write $y = kx$, where k is the variation constant. If distance varies directly as time, then $d = kt$, where k is the constant rate of speed.

Cruising Down the Road

Judy is on I-75 driving from Lexington to Knoxville and other points south. She is traveling late at night and there is very little traffic, so she sets her cruise control at 72 mph. How long will it take her to travel 200 miles?

1.   First, look at an input-output table for $d = kt$. The constant rate of speed is 72 mph, so $d = 72t$. What do d and t represent?

t	$d = 72t$
1 hour	72 miles
1.5 hour	108 miles
2 hour	144 miles
2.5 hour	180 miles
3 hour	216 miles

2.   Between what two values of t is the distance 200 miles?

3.   At what time is the distance closest to 200 miles?

The distance of 200 miles leads to the equation $200 = 72t$.

$$200 = 72t$$
$$\frac{200}{72} = t$$
$$t \approx 2.78 \text{ hours}$$

It takes Judy 2.78 hours, or approximately 2 hours and 47 minutes, to travel 200 miles. Explain how to change 2.78 hours to hours and minutes.

  Solve the equation numerically or graphically on your calculator. Do your answers agree?

Pressure in a Swimming Pool

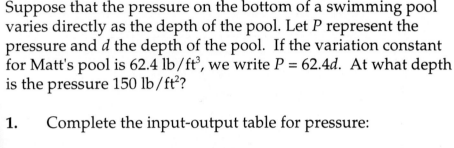

Suppose that the pressure on the bottom of a swimming pool varies directly as the depth of the pool. Let P represent the pressure and d the depth of the pool. If the variation constant for Matt's pool is 62.4 lb/ft^3, we write $P = 62.4d$. At what depth is the pressure 150 lb/ft^2?

1.   Complete the input-output table for pressure:

d	$P = 62.4d$
1 ft	
2 ft	
3 ft	
4 ft	
5 ft	

2.   At what depth is the pressure closest to 150 lb/ft^2?

3.   Write and solve an equation to find the depth (to the nearest tenth of a foot) where the pressure is 150 lb/ft^2.

4.   Solve the equation using your calculator. Do the answers agree?

You Try It

1. How long would it take Judy to travel 200 miles if her cruise control is set at 55 mph? At 70 mph?

2. In Matt's pool, at what depth is the pressure 75 lb/ft^2? 300 lb/ft^2?

3. Sales tax varies directly with the purchase price. The sales tax rate is the variation constant.

    a) The sales tax rate in McLean County is 7.5%. Complete the I-O table on the next page.

P (purchase price)	S = 0.075P (sales tax)
$10	
$50	
$100	
$1000	
$5000	
$10,000	

b) Laddie's Fireworks collected $350.17 in sales tax yesterday. How much did they sell?

How much money did Laddie earn? This might be your first thought after calculating how much he sold. To determine how much he earned, we need to understand markup and markdown. Problems involving price markup and markdown lead to percent equations.

Percent Equations

Before you begin studying percent equations you might want to review the arithmetic of percents. It is important to remember that percent means part of 100 and that arithmetic of percents is done with either decimals or fractions. So $30\% = \dfrac{30}{100} = 0.30$.

Marking Up Firecrackers

Laddie's Fireworks marks up their giant package 125%. If Laddie sells the packages for $79.99, how much did he pay for each package?

Obviously Laddie charges more than he paid in order to pay his expenses and make a profit. His markup percent is 125%.

We consider Laddie's cost for the item to be 100%. When we add the cost and markup percent, we get 100% + 125% = 225%.

This leads to the equation: List Price = 225% of Cost, or
$$L = 225\%C$$

Substitute the list price of $79.99 and solve for C. Round your answer to the nearest cent.

$$\$79.99 = 2.25C$$
$$\$35.55 = C$$

Laddie paid $35.55 for each giant package that he is selling for $79.99.

Suits for Less

Some percent equations involve markdowns. Three brothers bought suits on sale. Bill bought a new suit originally listed for $777 that had been discounted 25%. Russell bought a suit for $637 that was 33% off. Norman paid $450 for a suit listed for $675. Who got the best deal?

The firecracker example involved markup. The rate plus the cost, 100%, gave the list price. In markdown, we consider the list price to be 100%. The sale price rate will be found by subtracting the discount or markdown rate from 100%.

1. Let S be the sales price for Bill's suit. If his discount was 25%, the sale price would be $100\% - 25\% = 1.00 - 0.25 = 0.75 = 75\%$ of the list price.

    $S = 75\%$ of L
    $S = 0.75$ of 777
    $S = (0.75)(\$777)$
    $S = \quad \$582.75$

    The sale price for Bill's suit was $582.75.

2. Let the list price for Russell's suit be L. If he received a 33% discount, he paid 67%. Why?

    $\$637 = 67\%$ of L
    $\$637 = 67\%$ of L
    $\$637 = \quad 0.67L$
    $\dfrac{\$637}{0.67} = L$
    $\quad L = \$950.75$

    The list price for Russell's suit was $950.75.

3.   Let *P* be the percent of list paid by Norman.

$450 = p\%$ of L

$450 = P(\$675)$ *Remember that P represents the percent as a decimal.*

$\dfrac{\$450}{\$675} = P$

$P = 0.667$

Change *P* to a percent, 66.7%. Norman paid 66.7% of the list price. His discount was 33.3%. Why?

4.   Complete the following table to summarize the brothers' purchases:

Customer	List Price	Sales Price	Discount Rate	Discount Amount
Bill	$777		25%	
Russell		$637	33%	
Norman	$675	$450		

5.   Who saved the most money?

6.   Who received the largest discount percentage?

 7.   Who got the best deal? Write a report justifying your answer.

You Try It

Solve the following percent equations.

1. 40% of what number is 75?

2. What percent of 25 is 48?

3. $33\frac{1}{3}\%$ of 645 is what?

4. Kaye bought a pair of boots originally priced at $159.95 for $52.95. What discount rate did she receive?

5. Amie paid $139.96 for a set of tires which were on sale at 22% off. What was the list price for the set of tires?

6. Roscoe sold a customer a couch for $1150. If the couch was marked up 33%, what did Roscoe's Furniture pay for the couch?

Literal Equations

We have constructed input-output tables using formulas from geometry, science, and everyday life. Such formulas or equations with more than one variable are called **literal equations**. Often, we need to solve for one of the variables of a formula before we can construct an I-O table or work a problem.

These examples of formulas have been solved for one of the other variables. Later you might show the work yourself to fill in any gaps.

$$A = L \cdot W \quad \Rightarrow \quad L = \frac{A}{W}$$

$$\frac{1}{R} = \frac{1}{R_1} + \frac{1}{R_2} \quad \Rightarrow \quad R = \frac{R_1 R_2}{R_1 + R_2}$$

$$I = P \cdot R \cdot T \quad \Rightarrow \quad R = \frac{I}{P \cdot T}$$

We will now solve a literal equation and use it to complete an I-O table.

Temperature Revisited

In Unit 1, we used $F = \frac{9}{5}C + 32$ to construct a table of Fahrenheit temperatures for given Celsius temperatures. Suppose that you need to construct the following input-output table:

Fahrenheit °F	Celsius °C
−50	
0	
32	
100	
212	

Solve $F = \frac{9}{5}C + 32$ for C.

$$5(F) = 5\left(\frac{9}{5}C + 32\right)$$ *Multiply both sides by 5.*

$$5F = 9C + 160$$ *Simplify.*

$$5F - 160 = 9C$$ *Subtract 160 from both sides.*

$$\frac{5F - 160}{9} = C$$ *Divide both sides by 9.*

This is commonly written $C = \frac{5}{9}(F - 32)$.

 Now complete the input-output table.

Length of a Rectangle

We now return to the perimeter of a rectangle. Suppose the perimeter and width are given, and we need to solve for the length L. Recall that the perimeter of a rectangle is $P = 2L + 2W$.

The perimeter of a rectangle is 300 cm and the width is 45 cm. Find its length.

First, solve the perimeter formula below for L.

$$P = 2L + 2W$$
$$P - 2W = 2L \qquad \textit{Subtract 2W from each side.}$$
$$\frac{P - 2W}{2} = L \qquad \textit{Divide both sides by 2.}$$

Now substitute $P = 300$ cm and $W = 45$ cm.

$$L = \frac{P - 2W}{2} = \frac{300 \text{ cm} - 2(45 \text{ cm})}{2} = \frac{210 \text{ cm}}{2} = 105 \text{ cm}$$

The length of the rectangle is 105 cm.

Find the length of a rectangle with a perimeter of 350 inches and a width of 75 inches.

You Try It

Solve the following formulas for the indicated letter.

1. Area of a Rectangle: $A = L \cdot W$ for L

2. Area of a Triangle: $A = \frac{1}{2}bh$ for h

3. Interest: $I = PRT$ for R

4. Distance: $D = RT$ for R

5. Solve for *y*.

    a) $\qquad 4x + 2y = 10$

    b) $\qquad -3y + 6x = 20$

Inequalities

Sometimes problems involve *less than* ($<$), *more than* ($>$) , *at least* ($\geq$) or *at most* ($\leq$) rather than *equals*. We then need to solve an inequality rather than an equation.

Dee is collecting cola points and needs at least 750 to get a cooler. He has accumulated 567 points, and eleven weeks remain in the contest. On the average, how many cola points does he need to collect during each of the next eleven weeks to meet his goal?

If we let *a* be the average number of points collected each week, we write and solve the following equation:

$$567 + 11a = 750$$
$$11a = 183$$
$$a = \frac{183}{11}$$
$$a \approx 16.64$$

If Dee collects an average of 17 points per week, he will have sufficient points to earn his cooler.

Why is the following called an inequality?

$$567 + 11a \geq 750$$

The inequality $567 + 11a \geq 750$ will help us determine the number of points Dee needs to average per week to accumulate 750 or more cola points.

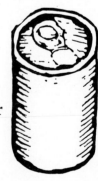

More or Less

To explore working with inequalities, begin with $5 < 10$. Add, subtract, multiply and divide each side of the inequality by 3. Examine the inequality to determine if it is still true. If false, change the inequality sign so it is true. Now repeat the process with -3, 4, and -4. What conclusions can you draw? Explain, in words, what you just learned.

We solve inequalities in much the same way as equations. We can add, subtract, multiply, and divide both sides by the same number. However, we reverse the order of the inequality when we multiply or divide by a negative quantity.

> If $a < b$, then
> $ac < bc$ if c is positive and
> $ac > bc$ if c is negative.
>
> If $a < b$, then
> $\frac{a}{c} < \frac{b}{c}$ if c is positive and
> $\frac{a}{c} > \frac{b}{c}$ if c is negative.

Solve the inequality:

$$567 + 11a \geq 750$$
$$11a \geq 183 \qquad \textit{Subtract 567 from each side.}$$
$$a \geq \frac{183}{11} \qquad \textit{Divide both sides by 11.}$$
$$a \geq 16.64 \qquad \textit{Simplify.}$$

Dee needs to collect an average of at least 16.64 cola points per week to reach his goal of at least 750 cola points. Notice the difference in the answers to the equation and the inequality for Dee's problem.

Gas for the Month

Ben enjoys driving. In addition to driving to the college every day, he runs errands for his mom and drives around with his friends. He has estimated that his tank holds enough gas to go another 250 miles, and he wants to keep a cushion of at least 50 miles through the first of next month in case his paycheck is late. If there are 22 days left in this month, how many miles can he drive

per day? Let *m* represent the number of miles he can drive per day. How many miles would he drive in 22 days? How many miles would be left after 22 days?

$$250 - 22m \geq 50 \qquad \text{Set up the inequality subtracting } 22m \text{ from } 250.$$

$$-22m \geq -200 \qquad \text{Subtract } 250 \text{ from each part.}$$

$$\frac{-22m}{-22} \leq \frac{-200}{-22} \qquad \text{Divide by } -22. \text{ The } \geq \text{ becomes } \leq.$$

$$m \leq 9.09 \text{ miles}$$

Ben can drive at most 9.09 miles per day and still have 50 or more miles left.

Compound Inequality

A **compound inequality** has two less than or greater than signs. For example, to earn a C in history, you need to score at least 70 but less than 80. This can be written as $70 \leq s < 80$.

The administration at the community college has decreed that the temperature of a classroom should be between 68°F and 72°F. Cindy has discovered that her thermostat measures 5° too hot. To calculate the range of the temperatures at which she should set her thermostat, she must solve the compound inequality $68 < T + 5 < 72$. To solve it, we perform the same operation on all *three parts* of the inequality.

$$68 < T + 5 < 72$$
$$68 - 5 < T < 72 - 5 \qquad \text{Subtract 5 from all three parts.}$$
$$63 < T < 67$$

Cindy should set her thermostat between 63 and 67 degrees Fahrenheit.

Solve another compound inequality.

$-3 < 2x - 4 < 7$	*The quantity $2x - 4$ lies between -3 and 7.*
$1 < 2x < 11$	*Add 4 to all three parts.*
$0.5 < x < 5.5$	*Divide all three parts by 2.*

The solution is all values of x between 0.5 and 5.5.

You Try It

Solve the following inequalities.

1. $3 + 4x < 8$

2. $22 - 11x < 14$

3. $12x + 7 < 2x + 40$

4. $4 - 5x < 11 - 3x$

5. $-3 < 2x + 4 < 5$

6. How many miles could Ben drive per day if he only needed to keep a reserve of 30 miles?

We have now solved both linear equations and inequalities. However, while each linear equation has had one solution, each inequality has many solutions. For example, Ben could drive at most 9.1 miles per day. Ben could also average 8, 7, 6, or fewer miles a day, or any number of miles a day less than 9.1. We often use a graph to help us visualize the infinite number of solutions.

Graphs of Inequalities

Algebraically, the solution of the inequality $x - 4 \leq -2$ is $x \leq 2$. Values that would make the original inequality true include 0, 1, 2, –3, –3.223 as well as an infinite number of others. To graph the solution $x \leq 2$ we draw a number line and place a solid circle slightly above the line at the point 2. The solid circle indicates that 2 is part of the graph. The solution includes all numbers less than or equal to 2, so we draw a line with an arrow to the left of 2.

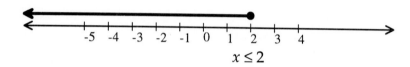

$$x \leq 2$$

 Solve and graph the following inequalities:

1. $-4x < 8$

Dividing both sides by –4, we find that $x > -2$. Values that make the original inequality true include –1, 0, 3 and others larger than –2.

Does –2 make the original inequality true? No, so –2 will not be part of the graph. To indicate that we are not including –2, we draw an open circle. The solution includes all numbers greater than –2. Therefore, draw a line with an arrow to the right of –2.

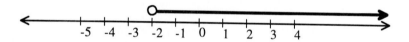

2. $-3 < 3(x - 1) \leq 5$

$$\frac{-3}{3} < x - 1 \leq \frac{5}{3}$$

$$0 < x \leq \frac{8}{3}$$

The values that make the inequality true include all numbers between 0 and $\frac{8}{3}$. It does not include 0 but does include $\frac{8}{3}$. On the line graph, the solid circle is about two-thirds past 2 since $\frac{8}{3} = 2\frac{2}{3}$.

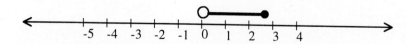

You Try It

Solve each of the following inequalities and graph the solution.

1. $x + 2 < 5$

2. $2a + 1 \geq 5$

3. $-2 \leq m - 3 < 2$

4. $3 \leq 7 - \dfrac{4}{5}b < 8$

Notations

Often a graph helps you understand the solution. Interval notation is an alternate way to write the solution. We have three ways to write the solution for an inequality.

Inequality Notation: $x \leq 2$

Line Graph:

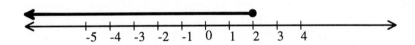

Interval Notation: $(-\infty, 2]$

Interval Notation

The table on the next page has both inequality and interval notation. Sketch the graph of each inequality in the row below it. Discuss with your group the use of parentheses and square brackets and the use of $-\infty$, which represents *negative infinity*, and ∞ or $+\infty$, which represent *positive infinity*. Answer the questions below and on the opposite page.

Inequality Notation	Meaning	Interval Notation
$x < 3$	x is less than 3	$(-\infty, 3)$
$x \leq 3$	x is less than or equal to 3	$(-\infty, 3]$
$x > -5$	x is greater than -5	$(-5, \infty)$
$x \geq -5$	x is greater than or equal to -5	$[-5, \infty)$
$-3 < x < 5$	x is between -3 and 5	$(-3, 5)$
$7 \leq x \leq 11$	x is between 7 and 11, inclusive	$[7, 11]$
$-5 < x \leq 11$	x is greater than -5 but less than or equal to 11	$(-5, 11]$

Use the table to answer the following:

1. When do you use square brackets?

2. When do you use parentheses?

3. When do you use ∞?

4. When is –∞ used?

When using interval notation, the interval always begins with the smallest number or the number farthest to the left on the number line.

You Try It

Complete the following table:

Inequality Notation	Meaning	Interval Notation
	x is less than –5	
		$(-\infty, -7]$
$x > 15$		
		$[5, \infty)$
$-3 < x \leq 5$		
		$[-3, 1]$
$-5 < x < 11$		

Summary

During your study of this unit, you have:

1. Solved fractional equations. We eliminated the fractions by multiplying both sides of the equation by the least common denominator;

2. Solved proportions by setting the ratio with the given information equal to the ratio of the unknown;

3. Solved equations arising from direct variation problems using $y = kx$, where k is the variation constant;

4. Solved percent equations involving sales tax, markdown, and markup;

5. Solved literal equations;

6. Expressed inequalities in standard and interval notation; and

7. Solved and graphed inequalities.

Unit 4 Problems for Practice

A. Solve the following equations algebraically and then either numerically or graphically. Compare your answers and rework any that are not essentially the same.

1. $3x - 8 = 15 + 8x$

2. $7 - 3c = 4c - 17$

3. $55y - 101 = 33y + 103$

4. $-5 + 11d = 16 - 5(d + 3)$

5. $2(3x - 5) = 3.5x + 7$

6. $-3(2b - 11) = 6(2b - 4)$

7. $-2(3c + 5) + 7(2c - 1) = -12$

8. $5(3k + 44) = -6 - (11k - 7)$

9. Twins, Jon and Jay, together weighed 163 ounces at birth. If Jon weighed 7 ounces more than Jay, how much did Jon weigh?

10. Worldwide, the percent P of smokers in the adult population t years after 1965 is given by $P = 43 - 0.64t$. According to this equation is the percentage of smokers increasing or decreasing? After how many years will the percent predicted be 11%?

B.   Determine the area and perimeter of each of the following geometric figures.

11.

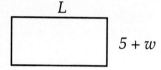

L

$5 + w$

12.

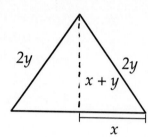

$2y$ $2y$

$x + y$

x

13.

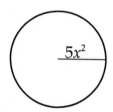

$5x^2$

14.

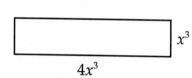

x^3

$4x^3$

15.

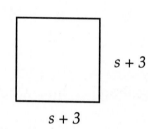

$s + 3$

$s + 3$

16.

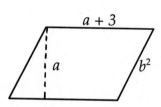

$a + 3$

a

b^2

17.

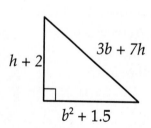

$3b + 7h$

$h + 2$

$b^2 + 1.5$

18.

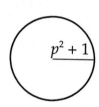

$p^2 + 1$

19.

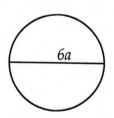

$6a$

20.   Write an expression that represents the area of the shaded region.

$3x + 3$

$x + 2$

3

3

C.   Simplify the following:

21. $21t + 13s - 5t - 8s$ **22.** $2a - 3b - 7 - 5a - 7b - 5$

23. $4(2x + 3) - 5(7 - 2x)$ **24.** $3(2w - r) - 4(5r - 3w)$

25. $4(21x + 13y) - 5(12x - 31) + 4(15 - 22y)$

26. $2x(5x - 3) - 7(5x - 3)$

27. $3a(4 - 3a) - 5(4 - 3a)$ **28.** $(3x - 2)(5x + 2)$

29. $(2a - 5b)(2a + 5b)$ **30.** $(x - 2)(3x + 5)$

D.   Write the following without negative or zero exponents:

31. $(-3x^2)(2x^3)$ **32.** $(3a^{-5})(5b^3)$

33. $(3x^{-2}y^2)(-5x^{-5}y^{-5})$ **34.** $(-3c^5d^0)(-5c^{-7}d^3)$

35. $(5x)^{-3}$ **36.** $5x^{-2}$

37. $(-7x)^0$ **38.** $-7x^0$

39. $5x^3y(z^0x^{-2}y^4 + 1)$ **40.** $-3z^{-3}(px^2 - 5z^3)$

41. $p^2q^{-3}(p - q + pq)$

42. $-3m^2n^{-2}(-2m^{-1} + n^2m^2 + 5n^{-4})$

43. $\dfrac{6x^2 - 3x}{3x}$

44. $\dfrac{40y^4 - 16y^3 + 8y^2}{8y^2}$

45. $\dfrac{8r^7 - 12r^5 + 6r^2}{2r^3}$

46. $\dfrac{100i^7 - 20i^5 + 40i^4}{40i^4}$

E. For each of the following:
   a) Estimate the answer;
   b) Evaluate using your calculator. Give answers in scientific notation, correct to two decimal places;
   c) Compare answers.

47. $(217,000)(777,000,000,000)$

48. $\dfrac{0.0000000001011}{727,000,000}$

49. $\dfrac{(5,210,000)(777,000,000)}{0.000003011}$

50. $(0.00050022)^3 (0.0000700044)^2$

F. Solve using scientific notation.

51. The French eat 2×10^7 frog legs per year. How many do they eat in 30 years?

52. The gray whale seen off the coast of California is the largest creature that has ever lived (larger even than the biggest dinosaurs). It weighs as much as 32 African elephants. An African elephant weighs 8 tons, or 16,000 pounds. How much does the average gray whale weigh?

53. Pluto's orbit around the sun is 5,900,000,000 kilometers. How far does it travel in 17 orbits? Write this in scientific notation.

54. Astronomers measure wavelengths in units called angstroms. One angstrom is 1×10^{-10} m. The wave length of red light is 6500 angstroms. Write the wavelength of red light in meters, in scientific notation.

55. The 1997 population of the United States is approximately 268 million. Write this in scientific notation.

56. The population of the Italian province of Tuscany is 3.5 million. This is approximately 6% of the entire population of Italy. In scientific notation, what is the population of Italy?

G. Solve the following fractional equations.

57. $\dfrac{2}{5} + \dfrac{x}{3} = 5$

58. $\dfrac{3}{2x} + \dfrac{1}{5} = \dfrac{5}{6}$

59. $\dfrac{1}{3} = \dfrac{1}{5} + \dfrac{1}{r}$

60. $\dfrac{3}{4} + \dfrac{2}{v} = \dfrac{7}{v} - \dfrac{2}{5}$

61. $\dfrac{7}{15} + \dfrac{x}{3} = \dfrac{5}{2} - \dfrac{x}{4}$

62. $\dfrac{1}{2} = \dfrac{1}{x} - \dfrac{5}{2x}$

63. Total resistance in a parallel circuit with three resistors is found using $\dfrac{1}{R} = \dfrac{1}{R_1} + \dfrac{1}{R_2} + \dfrac{1}{R_3}$. What is total resistance in a circuit with resistors of 3 ohms, 2 ohms, and 1.5 ohms? What is the third resistor in a circuit with total resistance of 2 ohms and resistors of 5 ohms and 6 ohms?

64. Two pipes are used to fill a large storage tank. Together, the flow from both pipes will fill the tank in three hours. If the flow from one pipe alone would fill the tank in five hours, how much time is required for the second pipe to fill the tank by itself?

65. The average cost to produce a quantity q of a product is calculated using $Avg\ cost = 1.27 + \dfrac{3700}{q}$. How many units must be produced so that the average cost is $2.75?

66. The average cost to produce mechanical pencils of a quantity q may be calculated using $C = 0.75 + \dfrac{2200}{q}$. How many pencils must be produced so that the average cost is $3.50?

H. Solve the following ratio equations:

67. $\dfrac{4}{x} = \dfrac{12}{35}$

68. $\dfrac{12}{35} = \dfrac{55}{y+1}$

69. $\dfrac{q-5}{4} = \dfrac{3(q+1)}{7}$

70. $\dfrac{2a-2}{3} = \dfrac{a+4}{7}$

71. Two cities are 11 inches apart on a map. If the scale of the map is 1 inch to 75 miles, how far apart are they?

72. A medicine has a dosage of 2.75 ml per 75 lb. How much medicine should a 250-pound person take? A 95-pound person?

73. One MB is 1×10^6 bytes. A 3.5 inch diskette holds 1.4 MB of data. How many bytes can 20 disks hold? How many disks are needed to hold 100 MB of data?

74. One inch equals 2.54 centimeters. How many centimeters are in 36 inches? How many inches are in 55 centimeters?

75. The average American eats 80 pounds of vegetables each year. If there are 2.5×10^8 Americans, approximately what is the number of pounds they consume in a year?

76. A light-year is approximately equivalent to 5.88×10^{12} miles. If the fastest space ship can travel at 77,000 miles per hour, how long would it take to get to Vega, which is 7.5 light-years from Earth?

I. Direct Variation

77. Sarah is driving home from college. If she sets the cruise control at 70 mph, how long will it take her to drive the 315 miles home?

78. Last Christmas, Sarah got a ticket for going 85 mph on the Pennsylvania Turnpike, where the speed limit is only 65 mph. She isn't taking any chances this trip since her mom gave her a really hard time about the $95 fine (not to mention the points on her license). She has set the cruise control at 68 mph. How far will she go in $2\frac{1}{2}$ hours?

79. Where should Sarah set her cruise control if she needs to get back to Princeton in 3 hours to make a seminar tonight, and she is 220 miles away?

80. David lives in Flagstaff, Arizona, and has a swimming pool. The pressure on the bottom of the pool varies directly with the depth, but the variation constant is only 57.5 lb/ft^3 since the altitude in Flagstaff is 7000 feet. What is the pressure at a depth of 4 feet?

81. At what depth is the pressure in David's pool 150 lb/ft^2?

82. Marquise found some money and is taking some of his friends to Graeter's for ice cream. If a dish of ice cream with one topping is $1.85, how much will it cost him to take 7 of his friends? If he has $30.00, how many friends can he take?

J. Solve the following:

83. 140% of what is 75?

84. What percent of 75 is 48?

85. $33\frac{1}{3}$% of 645 is what?

86. 300% of what is 60?

87.   Christopher bought a pair of shoes originally priced at $27.75 for $12.99. What discount rate did he receive?

88.   Debbie paid $52 for a set of speakers for her CD player. If the speakers were on sale at 22% off, what was their list price?

89.   Terry sold a customer a red sports car for $13,150. If the car was marked up 33%, what did Terry's Motors pay for it?

90.   The manager at Ski World decided that a markup rate of 36% is necessary to make a profit. What is the markup on a pair of skis that cost $162.95?

91.   The Green Thumbs nursery buys an ornamental fern for $36.75 and uses a mark-up rate of 46%. What is the list price?

92.   Santa's Tree Farm buys a Douglas fir for $22.50 and uses a mark-up rate of 52%. What is the list price?

K.   Solve the following literal equations for the indicated letter.

93. $5x - 6y = 30$ for y

94. $S = 4\pi r^2$ for π

95. $2x + 3y = 12$ for y

96. $S = \frac{1}{2}gt^2$ for g

97. $m = \dfrac{y-3}{x+2}$ for y

98. $a = \dfrac{d-40}{t-5}$ for t

99. Solve $A = \frac{1}{2}BH$ for B. Use your solution to complete the following I-O table.

Area	Height	Base
300 sq. in.	50 in	
500 sq. ft.	100 ft	
1000 m²	125 m	
750 sq. yd.	44 yd	

100. The area of a trapezoid that is 5 inches high is given by $A = \frac{5}{2}(B_1 + B_2)$, where B_1 and B_2 represent the lengths of the parallel sides. Solve for B_1, and use your solution to complete the following input-output table.

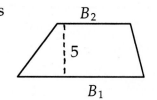

Area	B_2	B_1
27.5 sq. in.	5 in	
55 sq. in.	12 in	
25 sq. in.	8 in	
29.5 sq. in.	9.6 in	

101. Solve $P = 2l + 2w$ for l. Use your solution to complete the following input-output table.

P	w	l
39 cm	7 cm	
35.2 m	12.1 m	
17 mm	3 mm	
245.3 m	25 m	

102. The equation $D = 3.5t$ represents the distance you can travel in time t if you are traveling at a rate of 3.5 miles per hour. Solve the equation for t, and use your solution to complete the following input-output table.

D	t
39 miles	
35.2 miles	
17 miles	
245.3 miles	

L. Solve the following and express answers in both inequality and interval notation.

103. $3x - 5 < 6$ 104. $3 - 5x < 7$

105. $3x - 4 > 5 - 2x$ 106. $3(2x - 4) > 4(3x - 5)$

107. $-4 < 2x - 3 < 7$ 108. $8 > 3 - 4x > 3$

109. The temperature in your refrigerator should be between 20°F and 40°F.

   a) Write this as an inequality.

   b) Change this compound inequality to Celsius.

110. Lenora wants to build a rectangular pen for her ostriches, but space and fencing are limited. The pen must be 15 feet wide with a perimeter of at most 70 feet. What are the possible lengths she can make the pen?

111. The down payment for a house is at least 10% of the cost and the closing costs are $1100. If Rayanna and her husband have $12,100 in savings, what price homes can they consider purchasing?

112. An average between 80 and 89 earns Jim a grade of B in history. If his test scores are 72, 94, 83, and 70 and he has only one test left, what score can he make on the last test to earn a B?

M. Sketch the line graphs of the following:

113. $x > -2$

114. $[-8, -5]$

115. $-5 \leq x < 4$

116. $(-\infty, 5]$

117. $5 \leq x \leq 8$

118. $x < -5$

119. Graph the solution you found for **111**.

120. Graph the solution of problem **112**.

N.   Carol is designing an op-art quilt with triangles that she showed her group. One member of the group exclaimed that some of Carol's measurements can't possibly be correct. How did she know?

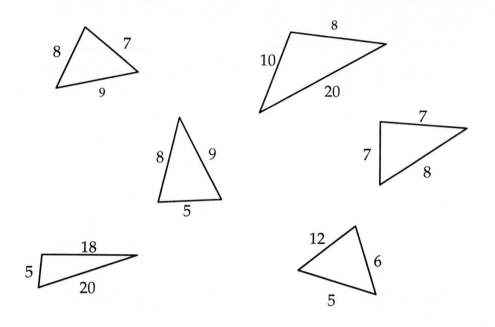

121.   Draw each triangle to scale using the scale $\frac{1}{4}$ inch represents 1 inch.

122.   If you couldn't draw any of the triangles, explain why not. What could you change to make it possible?

123.   Describe in words the relationship among the sides of a triangle needed. Using an inequality, write the relationship with a, b and c representing the sides of the triangle.

Slope and Equations of Lines

Upon successful completion of this unit you should be able to:

1. Find the slope and *y*-intercept of a line given its equation or its graph;

2. Sketch the graph of a line using the slope-intercept method;

3. Write the equation of a line given a graph;

4. Write the equation of a line given two points;

5. Write the equation of a line given the slope and a point; and

6. Write equations of vertical and horizontal lines.

Introduction

We used tables and the graphing calculator to graph lines in earlier units. In this unit, we will use the slope-intercept method to graph lines and to find equations of lines from graphs or from given points.

As you graphed a line you probably noticed the steepness of the graph and the point at which the line crossed the *y*-axis. These features will allow you to quickly graph any line and to write an equation to help you find other points. In the first example, we will consider the steepness of a line.

Changing Jobs

Suppose you tutor at the community center making $6 an hour. You see a notice on the college job board advertising a job at Ben's Car Wash paying $7 an hour. If you work ten or fewer hours each week, would it be worth changing jobs if the car wash is farther away and the job depends on the weather?

The pay for tutoring at the community center would be $y = 6t$ for *t* hours. For washing cars, the pay would be $y = 7t$ for *t* hours. Label the graphs so the solid line represents $y = 6t$ and the broken line represents $y = 7t$.

1. Which line is steeper?

2. Which line has the larger coefficient of *t* in its equation?

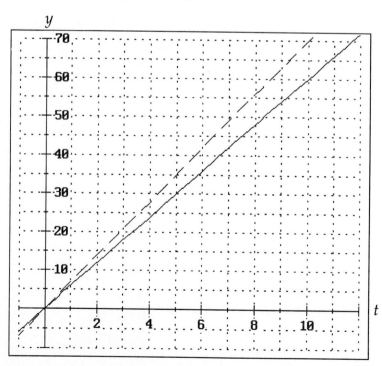

3. Complete the following input-output table to compare pay.

Hours t	Tutoring y = 6t	Car Wash y = 7t
2		
4		
6		
8		
10		
12		

4. Under what circumstances would you accept the car wash job and quit tutoring?

Review of Graphs of Lines Using Tables

Complete I-O tables and sketch the graphs of the following:

1. $y = 2x$

x	−3	0	3
y			

2. $y = 2x + 3$

x	−2	0	2
y			

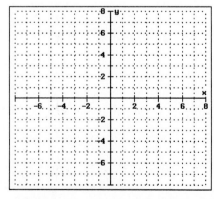

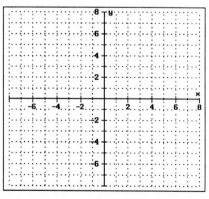

3. As you move left to right along the graph, compare the steepness of the graphs of $y = 2x$ and $y = 2x + 3$.

4. What is the connection between steepness and the coefficients of x in the equations of the lines?

5. $y = -3x$

x	-2	0	2
y			

6. $y = 2 - 3x$

x	-2	0	2
y			

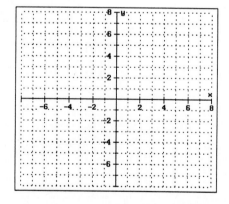

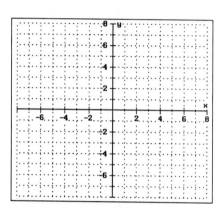

7.   Compare the steepness of $y = -3x$ and $y = 2 - 3x$. How does a negative coefficient of x affect the steepness of the line?

8.   Solve for y: $2x + 3y = 6$

$y =$ _____

9.   Solve for y: $2x - 3y = 6$

$y =$ _____

x	-3	0	3
y			

x	-3	0	3
y			

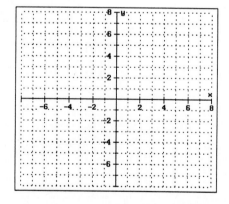

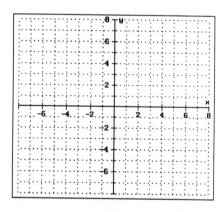

10.   Describe the connection between the steepness of each line and the coefficient of x in its equation.

Walk the Line

The next activity is designed to be completed after a demonstration of matching the graphs on the Calculator-Based Laboratory (CBL) or Calculator Based Ranger system (CBR).

For each of the following graphs, explain where a walker needs to begin, how fast a walker needs to walk, and the direction a walker needs to walk (away from or toward the motion detector).

1.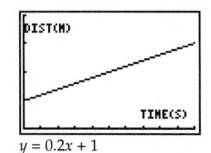

$y = 0.2x + 1$

2.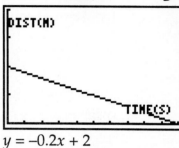

$y = -0.2x + 2$

3.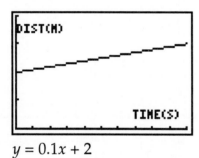

$y = 0.1x + 2$

4.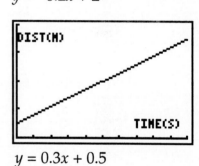

$y = 0.3x + 0.5$

5.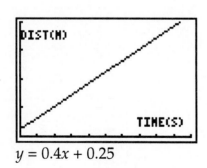

$y = 0.4x + 0.25$

6.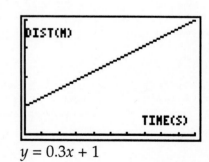

$y = 0.3x + 1$

7.   The walker would have to walk the fastest for which graph?

8.   For which graph(s) did the walker walk toward the motion detector? What is the sign of the coefficient of x for the equation?

9.   For which graph did the walker start the farthest away from the motion detector?

Slope of a Line

In the Walk the Line experiment, you saw that your speed determined the steepness of the resulting graph. The steepness of a graph is called its slope.

Suppose Sally hires a baby-sitter who charges $4 per hour. We write $y = 4x$ for the cost for x hours. Here is a table and graph of the cost depending on the number of hours. Label the axes.

x	0	1	2	3	4
y	0	4	8	12	16

What is the charge for:

5 hours   _____

3.5 hours   _____

1.75 hours   _____

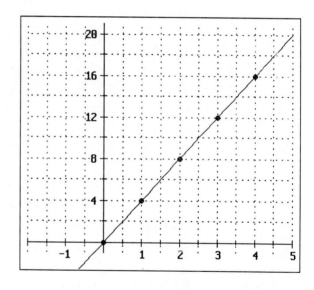

Notice the two triangles drawn on the baby-sitting graph. We call these **rise/run triangles**. A rise/run triangle uses only vertical and horizontal moves to move from one point on a line to another.

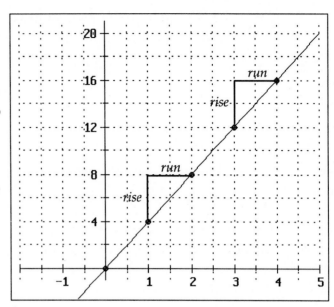

To move from the point (1, 4) to the point (2, 8), you rise or go up 4 units. You would then be at the point (1, 8). You then run or go to the right 1 unit. The $\frac{\text{rise}}{\text{run}}$ ratio is $\frac{4}{1}$. This $\frac{\text{rise}}{\text{run}}$ ratio is the slope of the line.

To move from (3, 12) to (4,16), you rise _____ units and run _____ units.

The $\frac{\text{rise}}{\text{run}}$ ratio is _____.

Draw rise/run triangles from points (1, 4) to (4, 16) and from (0, 0) to (3, 12). Calculate the slope for each triangle.

What do you observe about the value of the $\frac{\text{rise}}{\text{run}}$ ratio between any two points on the line?

The steepness of a line is called its **slope**. The slope is the $\frac{\text{rise}}{\text{run}}$ ratio or the ratio of the change in *y*-values to the change in *x*-values. The letter *m* is generally used for slope.

If two points of a line are (x_1, y_1) and (x_2, y_2), then

$$\text{slope} = m = \frac{\text{rise}}{\text{run}} = \frac{y_2 - y_1}{x_2 - x_1}$$

Let's return to the linear examples from the beginning of this unit and examine the slope of each.

Find the slope of the line indicated.

1. $y = 2x$

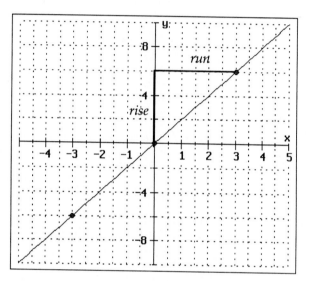

We've drawn a rise/run triangle from the point (0, 0) to the point (3, 6).

There is a rise of 6 and a run of 3.

$$m = \text{slope} = \frac{\text{rise}}{\text{run}} = \frac{6}{3} = 2.$$

Let $(x_1, y_1) = (0, 0)$ and $(x_2, y_2) = (3, 6)$.

$$m = \text{slope} = \frac{y_2 - y_1}{x_2 - x_1} = \frac{6 - 0}{3 - 0} = 2$$

The slope is 2 by each method. The coefficient of x is also 2 in the equation of this line.

Suppose we had drawn the rise/run triangle as shown below. Looking at the vertical change first, we must rise –6 units to move from (3, 6) to (3, 0). We then move to the left or run –3 units to move to the point (0, 0). The slope is
$$m = \frac{\text{rise}}{\text{run}} = \frac{-6}{-3} = 2.$$

Likewise, using the slope formula if we let $(x_1, y_1) = (3, 6)$ and $(x_2, y_2) = (0, 0)$, then

$$m = \frac{y_2 - y_1}{x_2 - x_1} = \frac{0 - 6}{0 - 3} = \frac{-6}{-3} = 2.$$

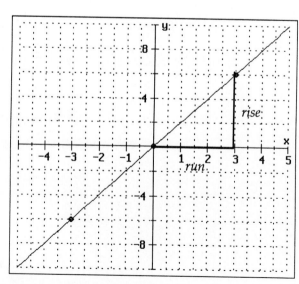

2. $y = -3x$

Draw a rise/run triangle from $(-3, 9)$ to $(3, -9)$.

The rise is _____, and

the run is _____.

m = slope = $\dfrac{\text{rise}}{\text{run}}$ = _____.

Let $(x_1, y_1) = (-3, 9)$ and $(x_2, y_2) = (3, -9)$.

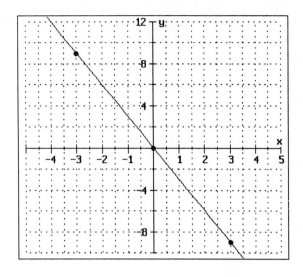

m = Slope = $\dfrac{y_2 - y_1}{x_2 - x_1} = \dfrac{-9 - 9}{3 - (-3)} = \dfrac{-18}{6} = -3.$

The slope is -3 by both methods. The coefficient of x is -3 in the equation of this line, also.

You Try It

Use the graphs you sketched on pages 177–178 to complete the following.

1. Draw rise/run triangles and use them to calculate the slope of $y = 2x + 3$.

2. Draw rise/run triangles and use them to calculate the slope of $y = 2 - 3x$.

3. Choose and label two points on the line of $2x + 3y = 6$. Calculate the slope using $m = \dfrac{y_2 - y_1}{x_2 - x_1}$.

4. Choose and label two points on the line of $2x - 3y = 6$. Calculate the slope using $m = \dfrac{y_2 - y_1}{x_2 - x_1}$

y-Intercepts

The **y-intercept** of a line is the point where the graph crosses or intersects the *y*-axis. At this point, the *x*-coordinate is zero. We frequently use the letter *b* for the *y*-coordinate.

> The *y*-intercept of a line is written as $(0, b)$.

Sally sometimes hires a different baby-sitter who charges $3 an hour plus a flat fee of $5 per sitting. The cost of *x* hours is $y = 3x + 5$.

Here is a graph of possible fees.

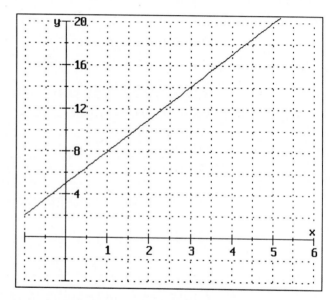

What is the charge for

5 hours _____

3.5 hours _____

1.75 hours _____

Where does the graph cross the vertical axis? $(x, y) =$ _____

This point where the graph crosses the vertical axis is called the *y*-intercept. In this application, what practical significance does this point have?

Draw several rise/run triangles for the graph. What is the slope of the line?

What is the practical significance of the slope?

1. Find the y-intercept of $y = 2x$.

    y-intercept $= (0, b) = (0, 0)$

    The line crosses the y-axis at the origin.

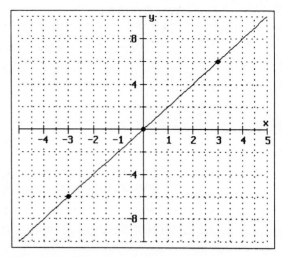

2. Use the graph below to find the y-intercept of $y = 2x + 3$.

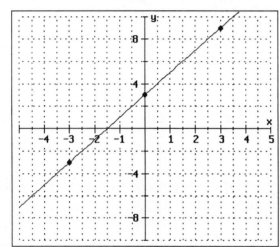

    y-intercept $= (0, b) = $ _____.

    This line crosses the y-axis at _____.

3. Compare and contrast the graphs of $y = 2x$ and $y = 2x + 3$. Can you look at the equation and write the y-intercept? How?

You Try It

Graph the following equations on your calculator, and trace to locate the y-intercepts.

	Equation	Slope	y-intercept
1.	$y = 2x + 3$		
2.	$y = 2 - 3x$		
3.	$2x + 3y = 6$		
4.	$2x - 3y = 6$		

Using the Slope and *y*-Intercept to Graph Lines

We can quickly draw the graph of a line if we know its *y*-intercept and slope. We plot the *y*-intercept and then use the slope to find additional points. We solve for *y* when the given equation is not in the form $y = mx + b$.

> The **slope-intercept form** of a line is $y = mx + b$, where $(0, b)$ is the *y*-intercept and m is the slope.

Let's use the slope and *y*-intercept to graph $y = 3x - 2$.

First, plot the *y*-intercept, $(0, b)$.

For $y = 3x - 2$, the *y*-intercept is $(0, -2)$. Plot this point.

Next we write the slope as a $\frac{\text{rise}}{\text{run}}$ ratio. The slope is 3, so $m = 3 = \frac{3}{1} = \frac{\text{rise}}{\text{run}}$.

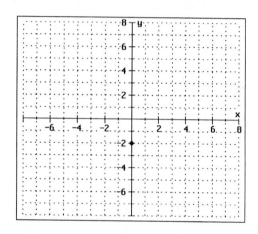

We plot a second point by moving from the *y*-intercept *rise* units up on the *y*-axis and *run* units over parallel to the *x*-axis.

For our example, from the *y*-intercept we go up three units and to the right one unit. This results in the point $(1, 1)$.

You can now find a third point by starting at the second point. Plot the third point by starting at $(1, 1)$, going up three units and to the right one unit. The third point is _____. Now sketch the line through the points.

You could use only two points to graph the line. However, a third point provides a check since any two points form a line. If the plotted points are not in a line, check your work. Once you have three points in a line, connect them.

We will use the slope-intercept method to graph several lines.

1. $y = -3x$ has a *y*-intercept of $(0, 0)$ and a slope $= \frac{\text{rise}}{\text{run}} = \frac{-3}{1}$. The slope has a rise of -3, which means that we go down three units on the *y*-axis. The run is $+1$, so go to the right $+1$ unit.

The first point to plot is the
y-intercept (0, 0).

From the y-intercept, go down 3
units and to the right one unit to
get a second point (1, –3).

From (1, –3), go down 3 units
and to the right one unit to get a
third point (2, –6).

Connect the points to draw the
line.

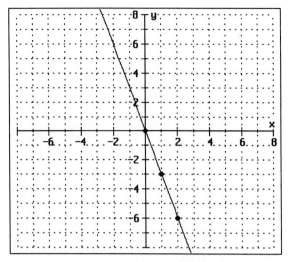

2. $y = \dfrac{-2}{3}x + 3$ has a y-intercept of _____ and a slope of $m = \dfrac{\text{rise}}{\text{run}} =$ _____.

On graph paper, plot the y-intercept.

Plot a second point, which is _____.

A third point is _____

Connect the points with a line.

Does your graph include the same points as this line?

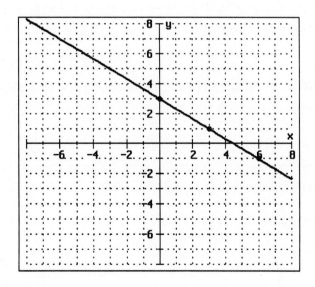

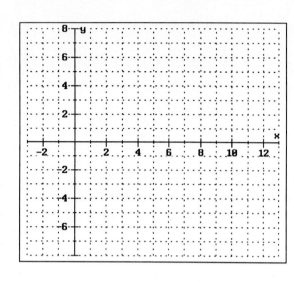

3. The graph of $y = \frac{2}{5}x - 3$ has a

   y-intercept of _____ and a

   slope of $m = \dfrac{\text{rise}}{\text{run}} = $_____.

   Plot the y-intercept.

   A second point is _____.

   A third point is _____.

   Connect the points with a line.

4. When you solve $3x + 5y = 15$ for

   y, you obtain $y = \dfrac{-3}{5}x + 3$.

   Plot the y-intercept of _____.

   The slope is $m = \dfrac{\text{rise}}{\text{run}} = $_____.

   A second point is _____.

   A third point is _____.

   Connect the points with a line.

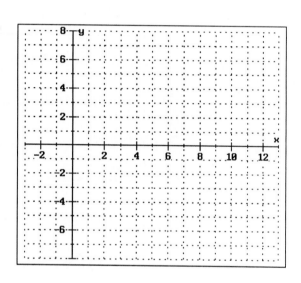

5. Look at the y-intercepts of the two lines in **3** and **4**. Where is the y-intercept when b is positive? When b is negative?

6. Consider the graphs in **3** and **4**. If you move from left to right, how does the line slant when m is positive? When m is negative?

You Try It

Use the slope-intercept method to find three points on the graph of each of the following, then connect the points with a line.

1. $y = \dfrac{-2}{3}x$

2. $y = \dfrac{3}{4}x - 2$

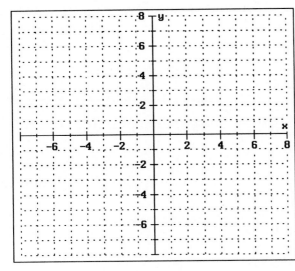

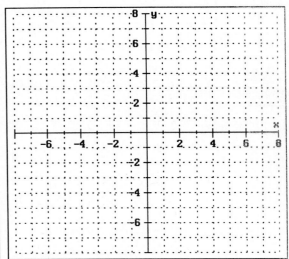

3. $4x + 3y = 12$

4. Graph the line with slope of $\dfrac{-2}{5}$ and a y-intercept of $(0, 3)$.

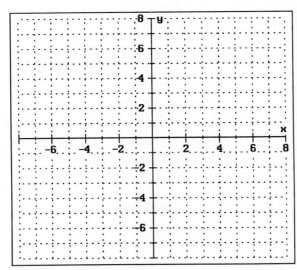

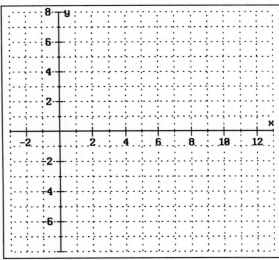

Writing Equations of Lines

We can use a line's slope and y-intercept to write its equation. Once you locate one point of a line, you can find the slope by drawing a rise/run triangle from that point to a second point. You can also calculate the slope using $m = \dfrac{y_2 - y_1}{x_2 - x_1}$.

Examine the following graphs in which you can easily read the y-intercepts from the graph. Determine the slope and y-intercept of the line to write the equation of the line.

Write the equation of each line in the form $y = mx + b$.

1.

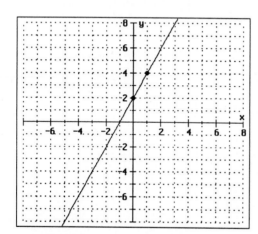

The y-intercept is $(0, b) = (0, 2)$.

To get to the point $(1, 4)$, we move up two units and to the right one unit.

Draw the rise/run triangle and label the rise and the run.

The slope is $m = \dfrac{\text{rise}}{\text{run}} = \dfrac{2}{1} = 2$.

The equation of any line is $y = mx + b$, where m is the slope and b is the y-intercept. The equation of this line is $y = 2x + 2$.

2.

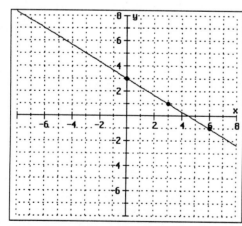

The y-intercept is $(0, b) = (0, 3)$.

To move to the point $(3, 1)$, we go down two units and to the right three units. Draw the rise/run triangle and label the rise and run.

The slope is $m = \dfrac{\text{rise}}{\text{run}} = \dfrac{-2}{3}$.

The equation of this line is $y = $ _____$x + $ _____.

3.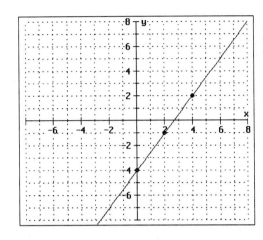

The y-intercept is $(0, b) =$ _____.

A second point is _____.

rise = _____ run = _____.

slope = $m =$ _____.

The equation is $y =$ _____.

4.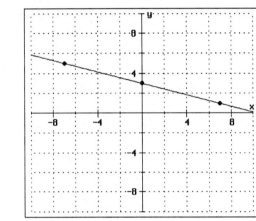

The y-intercept is $(0, b) =$ _____.

A second point is _____.

rise = _____ run = _____.

slope = $m =$ _____.

The equation is $y =$ _____.

This method works for a line if you already have the graph. You can also easily find the equation of a line if you know its slope and a point, or if you know two points on the line.

5. A line has a slope of 3 and a y-intercept of $(0, -5)$. Its equation is
$y =$ _____$x - 5$.

6. A line with points $(0, 5)$ and $(2, 7)$ has a slope of _____ and a
y-intercept of _____. Its equation is $y =$ _____.

You Try It

Find the equations of the following lines.

1.

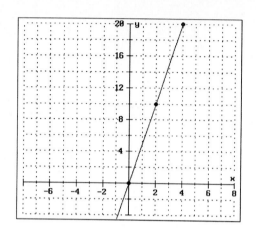

$y =$ _____

2.

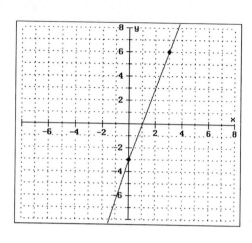

$y =$ _____

3.

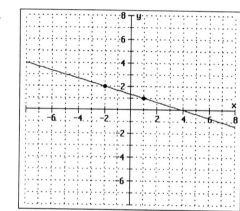

$y =$ _____

4. The line with points $(0, 2)$ and $(5, 5)$.

$m =$ _____

y-intercept is _____

$y =$ _____

5. A line with points $(-2, -14)$ and $(0, 4)$.

$y =$ _____

6. A line with a slope of $\dfrac{-13}{7}$ and a y-intercept of $\left(0, \dfrac{11}{3}\right)$.

$y =$ _____

Using Algebra to Determine *y*-Intercepts

Often the *y*-intercept is not easy to read from the graph. If so, we use algebra to determine *y*-intercept.

To find the equation of a line:

- Find the slope;
- Write the equation in the form $y = mx + b$ using the slope for *m*;
- Substitute one of the points for (x, y);
- Solve for *b*; and
- Write the equation using the slope for *m* and the *y*-intercept for *b*.

 Write the equation of the following lines.

1.

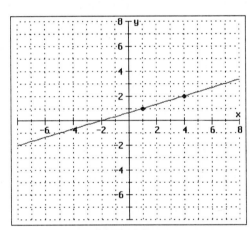

Draw a rise/run triangle from $(1, 1)$ to $(4, 2)$.

$$m = \frac{\text{rise}}{\text{run}} = \frac{1}{3}$$

$y = \frac{1}{3}x + b$ is the form of the equation of this line. Now find *b*.

To find *b*, substitute $(4, 2)$ for (x, y) and solve:

$$2 = \frac{1}{3} \cdot 4 + b$$
$$6 = 4 + 3b$$
$$2 = 3b$$
$$\frac{2}{3} = b$$

The equation of the line is $y = \frac{1}{3}x + \frac{2}{3}$. Verify that the value of *b* does not change if you use the point $(1, 1)$ instead of $(4, 2)$.

2.

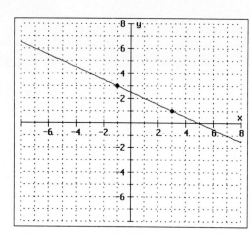

Draw a rise/run triangle from $(-1, 3)$ to $(3,1)$.

$$m = \frac{\text{rise}}{\text{run}} = \underline{\hspace{1.5cm}}$$

$y = \underline{\hspace{1cm}} x + b$

Choose one of the points, substitute for (x, y) and solve for $b = \underline{\hspace{1cm}}$.

The equation of the line is $\underline{\hspace{3cm}}$.

3. A line with points $(3, -5)$ and $(-7, 4)$ has a slope of $m = \dfrac{4 - (-5)}{-7 - 3} = \dfrac{9}{-10}$.

The equation of the line is $y = \dfrac{-9}{10}x + b$.

Substituting $(3, -5)$ for (x, y): $-5 = \dfrac{-9}{10}(3) + b$

$$-50 = -27 + 10b$$
$$-23 = 10b$$
$$\frac{-23}{10} = b$$

The equation of the line is $y = \underline{\hspace{3cm}}$

You Try It

Find the equation of each of the following lines.

1.

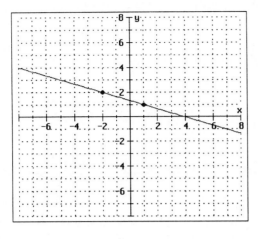

2.

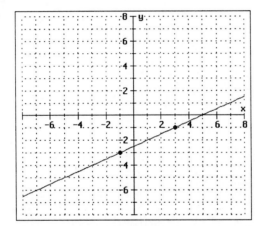

3. A line with points (1, −1) and (−5, 1).

4. A line with points (−1, −2) and (5, 2).

5. A line with points $\left(\dfrac{1}{3}, \dfrac{2}{5}\right)$ and $\left(\dfrac{-1}{5}, \dfrac{2}{3}\right)$.

Vertical and Horizontal Lines

Vertical lines do not fit the slope-intercept form, while horizontal lines have zero slopes. Let's look at examples of each and make observations about their equations.

Find the slope and equation of each of the following:

1.

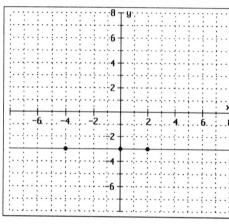

$(-4, -3)$ and $(2, -3)$ are points on the line.

$$m = \frac{\text{rise}}{\text{run}} = \frac{-3 - (-3)}{2 - (-4)} = \frac{0}{6} = 0$$

The y-intercept is $(0, -3)$.

The equation is $y = 0 \cdot x - 3$ or $y = -3$.

Look at the line again. What is the y-coordinate of each point on the line? How does this relate to the equation of the line?

2.

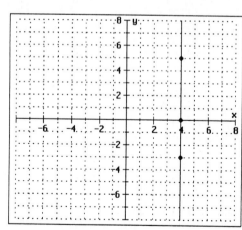

$(4, 5)$ and $(4, -3)$ are points of the line.

The slope $m = \frac{\text{rise}}{\text{run}} = \frac{-3 - 5}{4 - 4} = \frac{-8}{0}$ is undefined.

This line does not cross the y-axis. Therefore, the y-intercept does not exist.

At what point does the line cross the x-axis? _____ This point is called the x-intercept.

Examine the vertical line once again. What is the x-coordinate of each point of the line?

How does this relate to the equation of the line? The equation of the line is _____.

> **Horizontal lines** have a slope of zero, and their equations are of the form $y = b$, where $(0, b)$ is the y-intercept.
>
> **Vertical lines** have an undefined slope, and their equations are of the form $x = a$, where $(a, 0)$ is the x-intercept.

You Try It

Write the equation of each of the following lines.

1.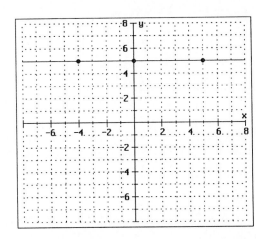

2.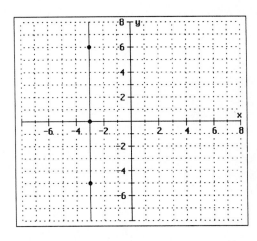

3. A line with points $(3, -17)$ and $(-4, -17)$.

4. A line with points $(-11, -7)$ and $(-11, 7)$.

5. Write the equation of a line parallel to the x-axis and 4 units above it.

6. Write the equation of a line parallel to and 4 units to the left of the y-axis.

Bouncing Ball

After a demonstration of the bouncing ball with the CBL, some students noticed that the relationship between the maximum heights the ball reached appeared to be linear. The screen below is from the experiment.

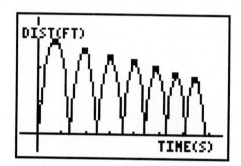

The instructor suggested the students drop a super ball from a variety of heights and measure the rebound height. The data in input-output table below represents the results of the second experiment.

Drop Height, inches	30	25	36	21	15
Rebound Height, inches	22.8	19	27.4	16	11.4

Your group should:

1. Plot the points.

2. Decide if the data represents a linear function and offer supporting documentation for your decision.

3. Calculate the slope of the line between each of the 10 pairs of points.

4. Write the equation of the line through the points (30, 22.8) and (36, 27.4). Also write the equation of the line through (36, 27.4) and (15, 11.4).

5. Choose two additional drop heights, and use your equation to predict the rebound heights. Construct a table to summarize your results, including the differences in predicted heights.

Summary

During your study of this unit, you:

1. Determined the slope of lines using rise/run triangles;

2. Calculated the slope of lines using the formula $m = \dfrac{y_2 - y_1}{x_2 - x_1}$;

3. Used the slope-intercept method to graph lines;

4. Wrote the equation of a line using the slope-intercept method. The slope-intercept form is $y = mx + b$, where $(0, b)$ is the y-intercept and m is the slope;

5. Used the equation $y = b$, where $(0, b)$ is the y-intercept and the slope is zero, to write the equation of a horizontal line; and

6. Wrote the equation of vertical lines. The equation of a vertical line has the form $x = a$, where $(a, 0)$ is the x-intercept and the slope is undefined.

Unit 5 Problems for Practice

A.   Solve:

1. $3 + 5b = -12$ 2. $-2(1 + 3g) = 10$

3. $5(2x - 4) = (x + 2)$ 4. $-3(x + 5) = 2(x - 7)$

5. $\dfrac{1}{2} + \dfrac{x}{4} = 7$ 6. $\dfrac{1}{5} + \dfrac{1}{x} = \dfrac{2}{3}$

7. $\dfrac{5}{k} = \dfrac{7}{28}$ 8. $\dfrac{a + 4}{7} = \dfrac{5}{2}$

9.   Alex drinks 3 cups of coffee each day. A cup of coffee has 180 mg of caffeine. Tea has about 35 mg of caffeine. How many cups of tea would he drink for the same amount of caffeine?

10.   How much caffeine does Tara drink during a day if she has 2 cups of coffee and 3 cups of tea?

11.   A Spanish dish, *paella*, calls for 75 ml of olive oil for 5 servings. All of Sara's family is coming to dinner, and she wants to make as much *paella* as possible, but she only has 200 ml of oil. She doesn't want to substitute—how many servings of *paella* can she make?

12.   Lauren wants to purchase a new bike that costs $180. She has saved $25 and has 5 weeks until she needs to begin training for the community race. On the average, how much should she save each week?

13.   One ounce is about 28.3 grams. A Spanish recipe calls for 150 g of pork. How many ounces do you need to buy for the recipe?

14.   On the average, Jose drinks two 12-ounce colas a day. A gallon contains 128 ounces. How many gallons of cola does he drink a year?

B.   Follow the directions on each exercise.

15.   Solve: $2x - 5 = -3x + 9$

16.   Solve: $4(2z - 15) = -3 - (7 - 4z)$

17.   Simplify: $2a(7 - 2a) - 3(11a - 5)$

18.   Simplify: $5(x - 2y) + 7(3y + x)$

19.   Simplify: $a^2 b^{-1}(b + a - ab^2)$

20.   Simplify: $-xy^{-2}(3x^{-1}y + 2xy^3 - 4y)$

21.   Solve: $\dfrac{1}{6} - \dfrac{x}{2} = 7$ 22.   Solve: $\dfrac{2}{7} - \dfrac{x}{3} = \dfrac{5x}{7} - \dfrac{1}{3}$

23.   Solve $7y - 2z = 11$ for z 24.   Solve $V = \frac{1}{3}\pi r^2 h$ for h

25.   Sarah is 4 centimeters taller than her younger brother Martin. Together they measure 319 cm. How tall is Martin?

26.   A medicine has a dosage of 1.25 ml per 50 lb. Omayra weighs 115 pounds. How much should she take?

27.   Terence needs a new backpack. The one he wants is regularly priced at $45, but is on sale for 12% off. How much will he have to pay for it?

28. Lupe needs at least a 90 average on her exams to make an A in algebra. Her exam scores so far are 79, 94, 95, and 87; there will be one more exam. To earn an A in the course, what does Lupe need to score on the last exam?

29. One mile is 1760 yards. How many yards are in 2.7 miles? How many miles are in 5000 yards?

30. The garden shop at the local grocery buys Christmas trees for $12 and sells them for 450% of their cost. How much do they charge for these Christmas trees?

C. Solve and graph the following inequalities.

31. $2x + 4 < 8$

32. $0.15x + 7 > -0.75x + 10$

33. $-5p + 2 \geq 14$

34. $2(-2x + 7) \leq 30 + 4x$

35. $\frac{1}{3}x - 2 < 7$

36. $\frac{2}{3}x - 4 > 5$

37. $0 \leq 2x - 5 < 7$

38. $5 < 0.2a + 3 \leq 7$

39. John is traveling to his parents' house. He estimates the trip will take between 2 hrs 10 min to 2 hrs 25 min. Write an estimate of the distance to his parents' house as an inequality if he travels at a constant speed of 60 mph.

40. Jennifer has made a New Year's resolution to save money. Her goal is to save at least $1000 this year. She has $125 toward her goal. On the average, what is the least amount she needs to save each month to meet her goal?

41.   Andrea and Chrissy want to buy a used golf cart to drive around
        the farm. A basic model costs $1200, and a deluxe model costs
        $1500. They have $250 now and would like to buy the cart in 7
        months. They want to know what is the least amount they need
        to save on the average each month for the basic and deluxe
        model.

42.   Jon is having a party and is expecting
        75 guests! He is calculating how
        many 6 packs of cola to buy. If each
        guest drinks at least 2 cans, how many
        6 packs does he need to buy?

43.   How many residents can squeeze into the dorm elevator if it will
        hold no more than 2500 pounds and the average student weighs
        160 pounds?

44.   Alexandra can't wait to ride the "big" rides at the amusement
        park. She must be at least 52" tall to ride the "big" roller coaster.
        Her dad says she is 48" tall. If she grows 1.5 inches per year, how
        long will it be before she rides the roller coaster?

D.   Determine the slope and *y*-intercept of each of the following:

45.

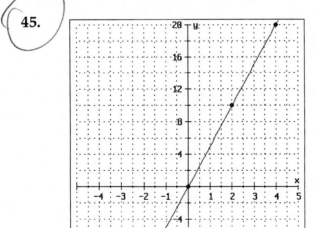

46.

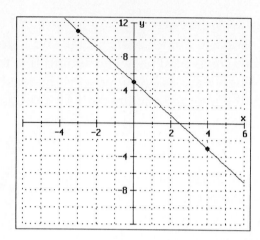

47.

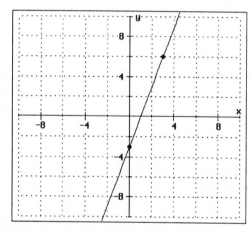

48.

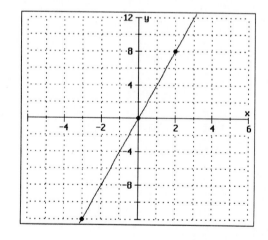

49.

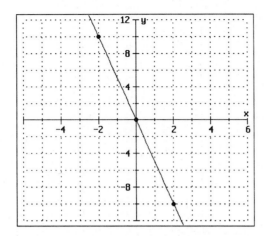

50.

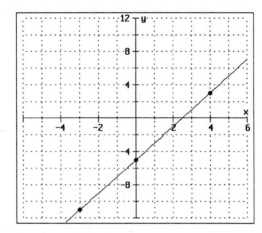

51.

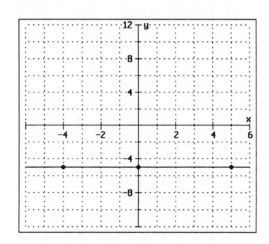

52.

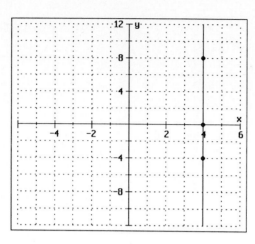

53.

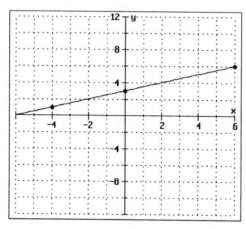

54.

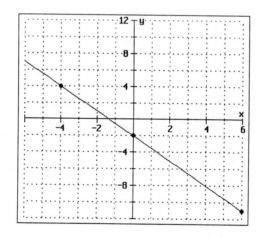

E.   Determine the slope for the line that passes through the following pairs of points.

55.   (5, –5) and (–5, 1)   **56.**   (5, –5) and (5, –1)

57.   (5, 10) and (10, –20)   **58.**   (–6, 1) and (1, –1)

59.   (1, 1) and (–4, 3)   **60.**   (7, –7) and (13, 7)

61. $\left(2, \dfrac{-8}{5}\right)$ and $\left(-4, \dfrac{-3}{2}\right)$ **62.**   (1, –1) and $\left(2, \dfrac{-8}{5}\right)$

63.   Does the table below represent a linear function? Hint: Are the slopes constant?

x	y
0	–12
2	–4
3	0
5	8
6	12

64.   Does the table below represent a linear function? Hint: Are the slopes constant?

x	y
–3	–14
2	4
4	8
5	11
7	16

65.   The bill from your lawyer for a one-fourth hour appointment was $22.50 and for a three-fourth hour appointment was $67.50. Calculate the slope in dollars per hour to determine her hourly charge.

66.   In 1980, the number of business trips over 100 miles was 97.1 million trips. By 1990, there were 155.6 million trips. Calculate the slope in business trips per year.

67.   The number of Americans who are 85 years old, or older, is rising linearly. In 1970, there were 1,408,000, and in 1990, there were 3,021,000 Americans who were 85 or older. Calculate the slope of the line through these points. What would be the units for your calculated slope?

68.   The line through points (a, b) and (c, d) has slope $m =$ _____, provided $a \neq c$.

F.   Record the slope, x-intercept and y-intercept, and graph the following:

69. $y = 3x - 5$ 70. $y = -5x + 3$

71. $4x - 5y = 20$ 72. $4x + 5y = 20$

73. $x = -3$ 74. $y = 5$

75. $y = \dfrac{1}{2}x + \dfrac{1}{2}$ 76. $y = \dfrac{-3}{5}x + \dfrac{2}{3}$

77.   Graph the line with a slope of $\dfrac{2}{3}$ that passes through $(1, 1)$.

78.   Graph the line with a slope of $\dfrac{-2}{5}$ that passes through $(-5, 3)$.

79.   Graph the line parallel to the x-axis and 4 units above it.

80.   Graph the line parallel to the y-axis and 5 units to the left of it.

81.   Graph the line through $(2, 5)$ and $(1, 2)$. Calculate the slope.

82.   Graph the line through (–3, –5) and (2, –2). Calculate the slope.

83.   Graph the line perpendicular to the *x*-axis that passes through (2, –4)

84.   Graph the line perpendicular to the *y*-axis that passes through (–2, 4)

85.   Graph the line that represents the cost of renting movies, if each movie costs $3.95 to rent. What is the *y*-intercept? What is the slope?

86.   Graph the line through the points (*Income, tax*) = ($7500, $550) and ($12,000, $820).

   a)   What is the slope? What does the slope represent?

   b)   Estimate the *y*-intercept. Explain what the *y*-intercept represents.

87.   Graph the line through the points (4 cookies, 150 calories) and (8 cookies, 300 calories). What is the slope of the line, and what does the slope represent?

88.   On the same graph as **87**, graph the line through the points (2 cookies, 90 calories) and (6 cookies, 270 calories). Calculate the slope. Compare the graphs from **87** and **88**.

G.   Refer to the graphs in **D**, then write the equation of each.

H.   Write the equation of each line in **E**.

Factoring and Quadratic Equations

Upon successful completion of this unit you should be able to:

1. Factor using the distributive law;

2. Factor differences of perfect squares;

3. Factor trinomials;

4. Solve quadratic equations algebraically, graphically, and numerically;

5. Use graphical and numerical methods to approximate factors;

6. Add, subtract, multiply, and divide rational expressions;

7. Use factoring to solve rational equations; and

8. Solve application problems.

Introduction

In this unit, we will study factoring and its uses in solving equations and simplifying fractions. We will also explore some methods we can use to approximate factors and to solve equations when we cannot factor expressions directly.

We can use an equation to solve the following problem:

The typical shuttle flight takes the vehicle away from land over the Atlantic Ocean. A pair of solid rocket boosters (SRBs) propel the shuttle to a speed of 3512 miles per hour. At approximately 2 minutes after launch, the SRBs separate from the shuttle. The SRBs are retrieved from the ocean to be reused for a later flight. The path of the SRBs from launch to landing can be approximated by $h = -0.009d^2 + 1.27d$, where h represents the height in miles and d represents the horizontal distance in miles of the SRB from the launch pad. The following graph shows the approximate path of the SRBs.

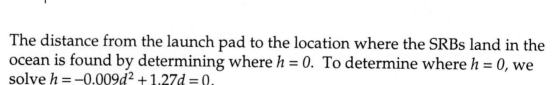

The distance from the launch pad to the location where the SRBs land in the ocean is found by determining where $h = 0$. To determine where $h = 0$, we solve $h = -0.009d^2 + 1.27d = 0$.

This type of equation is called **quadratic** because it is a polynomial and its largest exponent is 2. Later in the unit, we will use factoring and calculator methods to solve this equation. First we will review multiplication concepts from Unit 3.

Review of Multiplication of Expressions

We have used various methods to multiply algebraic expressions. Use an appropriate method to multiply each of the following:

1. $3(2x - 5)$

2. $(x + 3)(x - 3)$

3. $(x - 3y)(x + 4y)$

4. $3x(3x + 4)$

5. $3xy(x + y)$

6. $(2x + 5)(2x - 5)$

7. $(x + 1)(x - 5)$

8. $(c - 2d)(c - 5d)$

Factoring reverses multiplication. The **factors** in the above problems are the expressions to be multiplied. These were all in **factored form** before you multiplied.

Let's examine the factors of some of the products you found.

1. $3(2x - 5) = 6x - 15$

    The factors are 3 and $2x - 5$, and $3(2x - 5)$ is the factored form of the binomial $6x - 15$.

2. $(x + 3)(x - 3) = x^2 - 9$

    The factors are $x + 3$ and $x - 3$, and $(x + 3)(x - 3)$ is the factored form of the binomial $x^2 - 9$.

3. $(x - 3y)(x + 4y) = x^2 + xy - 12y^2$

    The factors are $x - 3y$ and $x + 4y$, and $(x - 3y)(x + 4y)$ is the factored form of the trinomial $x^2 + xy - 12y^2$.

You Try It

List the factors and the factored form for **4–8** on the previous page.

	Product	Factors	Factored Form
4.	$9x^2 + 12x$		
5.	$3x^2y + 3xy^2$		
6.	$4x^2 - 25$		
7.	$x^2 - 4x - 5$		
8.	$c^2 - 7cd + 10d^2$		

These products could be factored. However, if a polynomial will not factor, we say the polynomial is **prime**.

Factoring—Distributive Law

When we factor an expression, we write it as the product of its factors. Earlier we used the distributive law to multiply variable expressions. We can factor algebraic expressions with a common factor in each term using the distributive law.

> Distributive Law: $ab + ac = a(b + c)$

Factor each of the following:

1. $5x + 20$

    The expression $5x + 20$ has a greatest common factor of 5. We divide by 5 to find the other factor.

    $$\frac{5x}{5} + \frac{20}{5} = x + 4$$

    The factors are 5 and $x + 4$. The factored form of $5x + 20$ is $5(x + 4)$.

2. Factor $7x^2 - 49x$.

    a) What is the greatest common factor?

    b) What is the other factor?

    c) What is the expression in factored form?

3. Factor $8x^2y + 16xy^2 + 4xy$.

    a) What is the greatest common factor? What is the other factor?

    b) What is the expression in factored form?

4.   Factor $8x^2 + 15y^2$

The terms $8x^2$ and $15y^2$ do not have a common factor. Therefore, the expression cannot be factored. We say that $8x^2 + 15y^2$ is prime. Any expression that cannot be factored is prime.

You Try It

Factor the following. Label as prime those that cannot be factored.

1. $5c^2 - 30c$

2. $30b + 6$

3. $50d^2m^2 - 8m^2$

4. $11ab + 36cd$

5. $7xy^4 + 56x^2y^2 + 49x^3y$

6. $13a^2 - 17ab$

7. $2a^3b - 5a^2b^2 + 7ab^3$

Factoring—Difference of Perfect Squares

When we factor binomials, we first look for a common factor and use the distributive law in reverse. If there are no common factors, we examine each term to see if it has a square root. If it does, it may be an expression of the form $a^2 - b^2$, which factors as $a^2 - b^2 = (a - b)(a + b)$.

Perfect squares are numbers with square roots that are integers. Complete the following table:

x	1	2	3	4	5	6	7	8	9	10	11	12	13	14	15
x^2	1	4	9												

The numbers in the second row of the table represent the first 15 perfect squares.

You Try It

1. $\sqrt{225}$

2. $\sqrt{64}$

3. $\sqrt{196}$

4. $\sqrt{81}$

Difference of Perfect Squares: $a^2 - b^2 = (a - b)(a + b)$

 Factor the following:

1. $x^2 - 4$

   Since $x \cdot x = x^2$ and $2 \cdot 2 = 4$, the binomial $x^2 - 4$ is the difference of perfect squares, which factors into $(x - 2)(x + 2)$.

2. $4a^2 - 1 = ($ _____ $)($ _____ $)$

3. $4a^2 - 3$

   Even though $2a \cdot 2a = 4a^2$, 3 is not a perfect square. This is a not a difference of perfect squares and cannot be factored using the methods you've learned. Therefore, we say $4a^2 - 3$ is prime.

4. $7x^2 - 28$

   a) $7x^2 - 28 = 7($ _____ $)$

   b) Now use the difference of perfect squares to completely factor this expression.

Always look for common factors first. Often expressions that appear to be prime can be factored if you first use the distributive law to factor out the greatest common factor.

You Try It

Factor the following:

1. $x^2 - 49$

2. $36c^2 - 1$

3. $49t^2 - 4d^2$

4. $24z^2 - 121$

5. $5x^2 - 45$

6. $3t^3 - 27t$

Factoring Quadratic Trinomials

We will next explore methods for factoring quadratic trinomials. A trinomial of the form $ax^2 + bx + c$ is called quadratic if $a \neq 0$. The leading coefficient of the trinomial is a. The quadratic term ax^2 has coefficient a and is of degree 2. To factor these, we reverse the multiplying methods of Unit 3. When the leading coefficient a equals 1, we find factors of the constant c, whose sum is the middle term b.

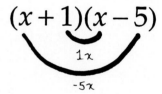

At the beginning of the unit, we multiplied $(x + 1)(x - 5)$ to get $x^2 - 4x - 5$. To factor $x^2 - 4x - 5$, we find factors of –5 with a sum of –4. The factors of –5 are $(-5)(1)$ and $(5)(-1)$. The factors –5 and 1 have a sum of –4. We write the factored form of $x^2 - 4x - 5$ as $(x - 5)(x + 1)$.

Factor the following:

1. $x^2 + 5x + 6$

The following table shows the factors of 6 and their sums:

Factors of 6	Sum
(1)(6)	$1 + 6 = 7$
(–1)(–6)	$-1 + (-6) = -7$
(2)(3)	$2 + 3 = 5$
(–2)(–3)	$-2 + (-3) = -5$

The factors of 6 that add to 5 are 2 and 3. So the trinomial $x^2 + 5x + 6$ is equal to $(x + 2)(x + 3)$ in factored form.

2. $b^2 - 4b - 12$

Complete the following table.

Factors of –12	Sum
(1)(–12)	
(–1)(12)	
(2)(–6)	
(–2)(6)	
(3)(–4)	
(–3)(4)	

Which factors add to –4?

$b^2 - 4b - 12 = (b - 6)(b + 2)$

3. $z^2 - 9z + 20$

Complete the table with the factors of 20. Both factors must be negative. Why?

Factors of 20	Sum

Which factors add to –9?

Write in factored form: $z^2 - 9z + 20 =$

4. $x^2 + 5x + 3$

The factors of 3 are:

Factors of 3	Sum

Neither pair of factors adds to 5. The trinomial $x^2 + 5x + 3$ is prime.

You Try It

Factor the following:

1. $a^2 - 2a - 15$ **2.** $b^2 - 8b + 15$

3. $v^2 + 2v - 15$ **4.** $c^2 + 3c - 10$

5. $x^2 + 3x + 5$ **6.** $r^2 + 5r + 6$

7. For the problems above, what patterns of signs in the factors have you noticed if the sign of the constant is positive?

8. For the problems above, what patterns of signs in the factors have you noticed if the sign of the constant is negative?

More on Factoring Trinomials

You can now factor quadratic trinomials in one variable with a leading coefficient of one. Let's examine trinomials that have other leading coefficients and trinomials with two variables.

Factor the following:

1. $x^2 - 2xy - 8y^2$

   $x^2 - 2xy - 8y^2 = (x - 4y)(x + 2y)$

   The factors of –8 that add to –2 are –4 and 2. However, the product needs to be $-8y^2$ and the sum $-2xy$. The y^2 in the third term of the trinomial is factored as $y \cdot y$. This leads to the last terms of –4y and 2y. If you multiply to check your answer, $(-4y)(x) = -4xy$ and $(x)(2y) = 2xy$ add to –2xy.

2.   Factor $a^2 + 12ab + 20b^2$.

   a)   What the factors of 20 that add to 12?

   b) $a^2 + 12ab + 20b^2 = (a + \underline{\hspace{1cm}}b)(a + \underline{\hspace{1cm}}b)$

3. $4x^2 - 12x - 16$

   a)   First, use the distributive law to factor out the monomial 4.

   b)   Complete the factorization.

4. $3m^2 - 6mn + 3n^2$

$$3m^2 - 6mn + 3n^2 = 3(m^2 - 2mn + n^2)$$
$$= 3(m - n)(m - n)$$

   We use the distributive law to factor out the 3. The other factor is a trinomial with a lead coefficient of 1, which we factor as before. The final answer could also be written as $3(m - n)^2$.

5.   Factor $3m^3 - 6m^2 + 3m$.

6.   Factor $3x^2 - 12x - 16$.

You Try It

Factor each of the following:

1. $x^2 - 5xy - 14y^2$

2. $2s^2 - 14st - 36t^2$

3. $3x^2 - 6x + 3$

4. $5m^2 + 30mn + 40n^2$

5. $5a^2 - 10ab - 30b^2$

6. $7x^3 - 10x^2 + 35x$

Solving Quadratic Equations by Factoring

Quadratic equations are of the form $ax^2 + bx + c = 0$, where $a \neq 0$. Why do we say $a \neq 0$?

Recall that the path of the solid rocket boosters (SRB) was approximated by $h = -0.009d^2 + 1.27d$, where h represents the height in miles and where d represents the horizontal distance in miles from the launch pad. We determine where the SRBs hit the ocean at height equal to zero by solving $-0.009d^2 + 1.27d = 0$.

How do we solve the equation?

First, factor $-0.009d^2 + 1.27d =$ _____.

What do we do with the factors? If $a \cdot b = 0$, what must be the value of either a or b?

Did you or your group conclude that either a or b or both had to be zero? The box below summarizes this fact.

> **Zero Factor Property:**
>
> If $ab = 0$, then $a = 0$ or $b = 0$ or both.

If $a \cdot b = 6$, what must be the value of either a or b? Several factors multiply to 6. Some values of a are 1, 2, 3, or 6. Possible values of b are also 1, 2, 3, or 6. We cannot conclude anything about the value of either a or b if $a \cdot b$ equals any number other than zero.

We can use this property to solve the SRBs equation:

$$-0.009d^2 + 1.27d = 0$$

$$d(-0.009d + 1.27) = 0 \qquad \textit{Factor out greatest common factor}$$

$$d = 0 \quad \text{or} \quad -0.009d + 1.27 = 0 \quad \textit{Set each factor equal to zero}$$

$$1.27 = 0.009d \qquad \textit{Solve each equation}$$

$$d = 0 \text{ miles or } d = \frac{1.27}{0.009} \approx 141 \text{ miles}$$

The SRBs are on the ground at 0 miles or at 141 miles. At the time of splashdown, the retrieval ships near the expected impact area quickly close in on the SRBs.

Use the zero factor property to solve two more problems.

1. The area of a rectangle of grass below is 5 square meters. The length and width are $x - 3$ and $x - 7$. What are the dimensions of the rectangle?

$A = 5 \text{ m}^2$	$x - 7$
$x - 3$	

$A = l \cdot w$	*Area formula*
$A = (x - 7)(x - 3) = 5$	*Area is 5 square meters*
$x^2 - 10x + 21 = 5$	*Multiply*
$x^2 - 10x + 16 = 0$	*All terms to one side of equation*
$(x - 8)(x - 2) = 0$	*Factor*
$x - 8 = 0$ or $x - 2 = 0$	*Use the zero factor property.*
$x = 8$ or $x = 2$	*Solve*

The solutions of the equation are 8 and 2. But what are the dimensions of the rectangle? Substitute 8 and 2 to get the length and width of the rectangle.

If $x = 8$, then $l = x - 3 = 8 - 3 = 5$ and $w = x - 7 = 8 - 7 = 1$. So the rectangle is 5 meters long by 1 meter wide.

If $x = 2$, then $l = x - 3 = 2 - 3 = -1$ and $w = x - 7 = 2 - 7 = -5$. Are these values reasonable? Why or why not?

The rectangle 5 meters long and 1 meter wide. Is its area 5 m²?

2. A right triangle has an area of 7 square yards. The height of the triangle is $x + 5$ and the base is x. What are the dimensions of the triangle?

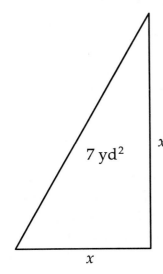

$$A = \frac{1}{2}B \cdot H \qquad \text{Area of a triangle}$$

$$\frac{1}{2}(x)(x+5) = 7 \qquad \text{Substitute}$$

$$x(x+5) = 14 \qquad \text{Multiply both sides by 2.}$$

$$x^2 + 5x = 14 \qquad \text{Multiply}$$

$$x^2 + 5x - 14 = 0 \qquad \text{Add } -14 \text{ to both sides}$$

$$(x+7)(x-2) = 0 \qquad \text{Factor}$$

$$x + 7 = 0 \text{ or } x - 2 = 0 \qquad \text{Zero factor property}$$

$$x = -7 \text{ or } x = 2$$

Is it reasonable for x to use the value of –7 in this problem. Why or why not?

Substitute $x = 2$ to find B and H. $B = 2 = 2$ yd and $H = 2 + 5 = 7$ yd. Is the area 7 yd²?

You Try It

Factor to solve the following:

1. $z^2 - 5z + 6 = 0$

2. $t^2 - 49 = 0$

3. $x^2 + 12 = -7x$

4. $r^2 + 5r = 6$

5. Find the length and width of a rectangle that has an area is 135 square inches if the width is given as $2x + 3$ and the length as $2x - 3$.

6. Find the base and height of a triangle with an area of 17.5 square feet if the base is given as $x - 3$ and the height as $x - 5$.

More Quadratic Equations

Examine the following equations. How do they differ from the ones in the last section?

1. $2v^2 + 4v = 30$

2. $4c^2 = 121$

3. $6d^2 + 7d - 3 = 0$

4. $4x + 10 + 2x^2 = 0$

All of these equations have a leading coefficient greater than one. Let's solve the equations and look for other similarities and any differences. To write a quadratic equation in **standard form**, you add or subtract terms as needed to obtain a zero on one side of the equation. Next, arrange the terms so that the term with the largest exponent is first, followed by the next lower exponent, followed by the term with the lowest exponent. The constant term could be written with a factor of x^0. In standard form, a quadratic equation is $ax^2 + bx + c = 0$.

 Solve.

1. $2v^2 + 4v = 30$

$2v^2 + 4v - 30 = 0$ *Write equation in standard form*

$2(v^2 + 2v - 15) = 0$ *Factor out the greatest common factor*

$2(v + 5)(v - 3) = 0$ *Factor trinomial*

$(v + 5)(v - 3) = 0$ *Divide both sides by 2*

$v + 5 = 0$ or $v - 3 = 0$ *Use zero factor property*

$v = -5$ or $v = 3$ *Solve*

2. $4c^2 = 121$

$4c^2 - 121 = 0$ *Write equation in standard form*

$(2c + 11)(2c - 11) = 0$ *Factor using difference of perfect squares*

$2c + 11 = 0$ or $2c - 11 = 0$ *Use zero factor property*

$c = \dfrac{-11}{2}$ or $c = \dfrac{11}{2}$ *Solve*

The next two examples are different in another way—neither can be factored using methods you've learned in this unit.

3. $6d^2 + 7d - 3 = 0$

There is no common factor for the terms of $6d^2 + 7d - 3$. We cannot factor the trinomial at this point, and therefore we cannot solve the equation by factoring.

4. $4x + 10 + 2x^2 = 0$

$2x^2 + 4x + 10 = 0$ *Write equation in standard form.*

$2(x^2 + 2x + 5) = 0$ *Factor out the greatest common factor.*

We cannot factor $x^2 + 2x + 5$ because there are no factors of 5 that add to 2. Thus, we cannot solve the equation by factoring.

Examples **3** and **4** both contain expressions that could not be factored using the methods developed in the unit.

Later, we will use graphical and numerical methods to solve such equations, if they have real solutions. Before we look at calculator methods of solving equations, let's look at one more equation.

 Solve $v^3 + 2v^2 = 15v$.

This is not a quadratic equation, since the monomial with the highest power is degree 3. We can obtain a quadratic equation if we first remove the common factor v.

$$v^3 + 2v^2 = 15v$$

$$v^3 + 2v^2 - 15v = 0$$

$$v(v^2 + 2v - 15) = 0$$

$$v(v + 5)(v - 3) = 0$$

$$v = 0 \text{ or } v + 5 = 0 \text{ or } v - 3 = 0$$

$$v = 0, v = -5, \text{ or } v = 3$$

Notice that there were three variable factors. This led to three equations and three solutions.

You Try It

Solve by factoring.

1. $2a^2 - 32 = 0$

2. $4x^4 = 169x^2$

3. $y^3 + 9y^2 + 18y = 0$

4. $5x^2 + 60 = -35x$

Summary of Solutions

We have used the zero factor property to find the solutions of quadratic equations. We wrote the equation in *standard form*, $ax^2 + bx + c = 0$, factored, and set each factor equal to zero. While you are completing the next activity watch for any patterns you observe relating factors and solutions. The *x*-values of the window suggested allow you to trace using values that change by tenths. The suggested *y*-values allow you to observe more of the graph, which is called a parabola.

Zeros

Clear all equations from Y= on your calculator, and set the window to $[-4.7, 4.7]_x$ and $[-12, 12]_y$.

```
WINDOW
 Xmin=-4.7
 Xmax=4.7
 Xscl=1
 Ymin=-12
 Ymax=12
 Yscl=1
 Xres=1
```

1. Complete the table below. **(a)** Graph and trace to determine the x-intercepts, which is the x-value where y is zero. **(b)** Factor the expression on the right-hand side of each quadratic equation, and set it equal to zero. **(c)** Use the zero factor property to solve the equation from the column with the factored form.

	x-Intercepts	Factored Form	Algebraic Solutions
$y = x^2 - 2x - 3$		$(x-3)(x+1) = 0$	
$y = x^2 - 4$			
$y = x^2 + 5x + 6$			
$y = -x^2 - 2x$			
$y = x^2 + 2x + 1$			
$y = 36 + 6x^2 - 30x$			
$y = 4x^2 - 49$			
$y = -9 + x^2$			

2. What is the connection between the x-intercept and the algebraic solution?

3. What is the connection between the factored form and the algebraic solution?

In the last activity, you graphed $y = x^2 - 2x - 3$ and found the x-intercepts $(-1, 0)$ and $(3, 0)$. You then factored the expression $x^2 - 2x - 3$ as $(x-3)(x+1)$ and set it equal to zero. Finally, you solved $(x-3)(x+1) = 0$ using the zero factor property to obtain two algebraic solutions, $x = 3$ and $x = -1$. What is the connection?

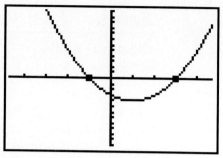

The factored form is $y = (x-3)(x+1)$. The output is zero when the input is either 3 or -1. The input 3 causes the factor $x - 3$ to be zero, and the input -1 causes the factor $x + 1$ to be zero. When either factor is zero, the product is zero. If an input causes an output to be zero, the input is called a **zero**.

If you graph an equation, an x-intercept is any point where the value of y is zero. Therefore, the x-value of the x-intercept is the zero. Solving the equation $x^2 - 2x - 3 = 0$ is the same as finding the zeros of $y = x^2 - 2x - 3$.

The table below summarizes the connections between factors and zeros.

Equation	Standard Form	Factored Form	Zeros
$2v^2 + 2v = 24$	$2v^2 + 2v - 24 = 0$	$2(v+4)(v-3) = 0$	-4 or 3
$4c^2 = 81$	$4c^2 - 81 = 0$	$(2c-9)(2c+9) = 0$	4.5 or -4.5
$6z^2 - 30z + 36 = 0$	$6z^2 - 30z + 36 = 0$	$6(z-3)(z-2) = 0$	3 or 2
$6d^2 + 7d = -3$	$6d^2 + 7d + 3 = 0$	prime	none
$4x + 10 = -2x^2$	$2x^2 + 4x + 10 = 0$	$2(x^2 + 2x + 5) = 0$	none
$v^3 + v^2 = 6v$	$v^3 + v^2 - 6v = 0$	$v(v+3)(v-2) = 0$	$0, -3, 2$

Use your graphing calculator to verify the zeros given. Enter the expression on the left-hand side of the standard form in $\boxed{\text{Y=}}$, and then use $\boxed{\text{TRACE}}$ to find the zeros.

Solving Quadratic Equations Graphically and Numerically

Earlier, we encountered a rectangle with area of 5 square meters, length of $x - 3$, and width of $x - 7$. The solutions to $x^2 - 10x + 16 = 0$ were $x = 8$ or $x = 2$. We rejected $x = 2$, because it led to a negative length. The solution $x = 8$ led to a length of 5 m and width of 1 m.

Graphically

To find the solution on the calculator, enter $x^2 - 10x + 16$ in **Y1**, and select the window shown.

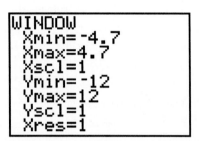

Press GRAPH. Your screen should match this one. How many zeros (points where the graph crosses the x-axis) do you observe? How many zeros do you expect?

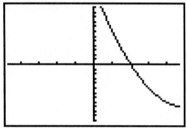

Press TRACE. Press the right arrow, ▶, until you can locate the x-value where $y = 0$. Record this value. If you continue pressing the right arrow, you will notice that the screen will change so that you can see more of the graph. Continue pressing ▶ until you find the second zero. Record the x-value where $y = 0$.

The points $(2, 0)$ and $(8, 0)$ are the x-intercepts. The zeros $x = 2$ and $x = 8$ agree with those we found by factoring. Recall that $x = 2$ led to a negative length, which we rejected. We substitute $x = 8$ to find $L = 5$ meters and $W = 1$ meter.

Solve by graphing $6d^2 + 7d - 3 = 0$.

We cannot solve this equation by factoring using methods developed in this unit, because we cannot factor 6 from each term of $6d^2 + 7d - 3$. Expressions like this can be factored, but finding the factors is much more complicated.

To solve $6x^2 + 7x - 3 = 0$ on the calculator, enter $6x^2 + 7x - 3$ in **Y1** and select the window $[-4.7, 4.7]_x$ and $[-12, 12]_y$. Graph and observe the zeros—the points where the graph crosses the x-axis. Trace to locate the zeros, if possible.

One zero was –1.5. However, by tracing the graph in this window, we cannot find the exact value of the second zero. Between what two values does the zero lie? Change the window to $[-3,3]_x$ and $[-3,1]_y$ to get a better look at the zeros. Use ⊡TRACE⊡ to find the zeros of the equation. Your calculator window should look like the ones below:

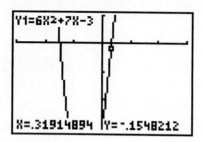
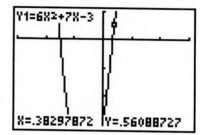

By tracing we find that the point for the second zero will have an x-coordinate, $0.32 < x < 0.38$. We can now use a table to get a closer approximation.

Numerically

Press ⊡2nd⊡⊡WINDOW⊡ and enter 0.32 as **TblStart** and 0.001 as **ΔTbl**. Press ⊡2nd⊡⊡GRAPH⊡ to see the table. Arrow down until **Y1** changes from negative to positive. The solution appears to be between 0.333 and 0.334 on the table since the value of **Y1** is close to zero.

To confirm the other solution numerically, enter –1.5 as **TblStart** and 0.001 as **ΔTbl**.

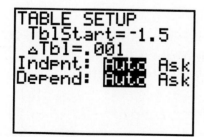

The solutions to $6x^2 + 7x - 3 = 0$ are $x = -1.5$ and $x \approx 0.33$.

What are the factors?

To get the solutions or zeros from factors, we set each factor equal to zero and solve. Notice the connection in the following table:

Zero	Equation	Factor
$x = -1.5$	$x + 1.5 = 0$	$x + 1.5$
$x \approx 0.33$	$x - 0.33 \approx 0$	$x - 0.33$

Is $6x^2 + 7x - 3 \approx 6(x + 1.5)(x - 0.33)$? We need to multiply by six since we know that $x \cdot x = x^2$, not $6x^2$.

Multiply to Verify: $6(x + 1.5)(x - .33) = 6(x^2 + 1.17x - 0.495)$
$$= 6x^2 + 7.02x - 2.97$$

When each coefficient is rounded to the nearest integer, this expression is $6x^2 + 7x - 3$.

 Solve $15a^2 - 6 = a$

Write in standard form: $15a^2 - a - 6 = 0$

To find the solution on the calculator, enter $15x^2 - x - 6$ in **Y1**, press $\boxed{\text{ZOOM}}\boxed{4}$ or set the window to $[-4.7, 4.7]_x$ and $[-3.1, 3.1]_y$ and graph **Y1**. Using $\boxed{\text{ZOOM}}\boxed{4}$ or this window graphs **Y1** so that the vertical distance is the same as the horizontal distance. Now, when you trace each x-value is 0.1 apart. Trace to observe the zeros.

Record the x-values that the zeros lie between: _____ and _____ or _____ and _____.

To get a closer look at the zeros, you might zoom in. Press $\boxed{\text{ZOOM}}\boxed{2}$. At this point the calculator waits for you to decide where you want the center to be. Move the pointer to $(0.7, 0)$. Press $\boxed{\text{ENTER}}$.

Trace to find the zeros of the equation. Your calculator windows might look these.

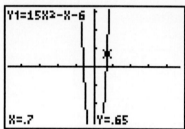 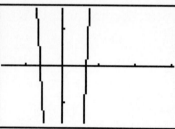 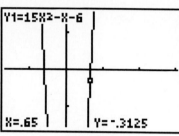

Before Zooming In   Centered on the second   Tracing

$x = -0.6$ and $x = 0.66$ are the apparent zeros or x-intercepts. Use a table to get a closer approximation. Give the zeros correct to two decimal points.

Use the zeros to factor: $15a^2 - a - 6 =$ (   )(   )

 Solve $x^2 + 2x + 2 = 0$ graphically.

When we graph $y = x^2 + 2x + 2$ on the calculator, we observe that the parabola does not intersect the x-axis. Therefore, $y = x^2 + 2x + 2$ has no zeros, and $x^2 + 2x + 2 = 0$ has no real solution.

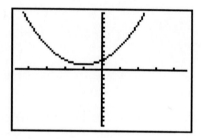

You Try It

Begin with the window $[-4.7, 4.7]_x$ and $[-12, 12]_y$ to solve graphically. Use the approximate graphical solution to solve numerically correct to 2 decimal places.

1. $2x^2 + 4x - 10 = 0$

2. $2x^2 - 9 = 0$

3. $3z^2 + 18 = 15z$

4. $6x^3 + 7x^2 = 3x$

5. $x^2 = -1$ **6.** $5x^2 - 3x + 8 = 0$

7. Approximate the factors of $2x^2 + 4x - 10$.

8. Approximate the factors of $3z^2 - 15z + 18$.

Algebraic Fractions

The following problems require factoring, finding the least common multiple (LCM) of the denominators, or finding the reciprocal of the divisor.

Perform the indicated operations.

1. $\dfrac{2}{3x} + \dfrac{5}{x^2}$

The LCM of $3x$ and x^2 is $3x^2$.

$$\frac{2}{3x} + \frac{5}{x^2} = \frac{2x}{3x^2} + \frac{15}{3x^2}$$
$$= \frac{2x + 15}{3x^2}$$

We multiplied the first fraction by $\dfrac{x}{x}$ and the second fraction by $\dfrac{3}{3}$. We then wrote the numerators over the common denominator.

2. $\dfrac{7x+2}{x^2+x} - \dfrac{5}{x+1} = \dfrac{7x+2}{x(x+1)} - \dfrac{5}{x+1}$

$$= \frac{7x+2}{x^2+x} - \frac{5x}{(x+1)x}$$
$$= \frac{7x+2-5x}{x^2+x}$$
$$= \frac{2x+2}{x^2+x}$$

Again, we found the LCM and multiplied the numerators by the appropriate factors. We combined like terms to get the answer.

3. $$\frac{7x}{x^2-3x} \cdot \frac{x-3}{5x} = \frac{7x}{x(x-3)} \cdot \frac{x-3}{5x} = \frac{7x \cdot (x-3)}{x(x-3) \cdot 5x} = \frac{7}{5x}$$

Notice the factors of $x-3$ and x in both the numerator and denominator. The product $x(x-3)$ divided by $x(x-3)$ is equal to 1.

4. Complete: $$\frac{x^2-9}{x^2+3x} \div \frac{x^2-2x-3}{x^2} = \frac{(x+3)(x-3)}{x(x+3)} \div \frac{(x-3)(x+1)}{x^2}$$

You Try It

Perform the indicated operations:

1. $\dfrac{5}{4x^2} - \dfrac{2}{3x}$

2. $\dfrac{7}{x-2} + \dfrac{x+1}{x^2-2x}$

3. $\dfrac{x^2-9}{5x} \cdot \dfrac{7x^2}{x+3}$

4. $\dfrac{x^2+5x}{x^2-25} \div \dfrac{x^2}{x^2-6x+5}$

In Unit 4, we found the solution of a fractional equation by multiplying both sides of the equation by the least common multiple. Later we will study some equations that require factoring to find the LCM.

Fractional Equations

Similar objects have the same shape, but not
necessarily the same size. Basketballs are similar
to baseballs, but a football is not similar to either.
Furniture in a design model is similar to the
furniture to be manufactured for an office.
Similar objects have corresponding parts, and the
relationship between the sizes of corresponding
parts can be written as a ratio. Each ratio will be
the same.

In the similar triangles below, which are
corresponding sides?

If the measurements are as below, we can write two ratios where the first ratio
arises from the larger triangle and the second from the smaller.

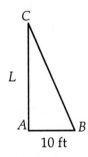

$$\frac{L}{10 \text{ feet}} = \frac{2 \text{ inch}}{0.4 \text{ inch}}$$

Now solve the equation to determine *L*.

More complicated relationships will require the methods you've learned in this
unit, as in the following example.

The triangles below are similar triangles. Determine the measure of each of the unknown sides of the triangles.

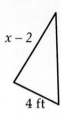

$$\frac{x-2}{4} = \frac{6}{x+3}$$ *Set up the ratios*

$$\frac{4(x+3)(x-2)}{4} = \frac{4(x+3)6}{x+3}$$ *Multiply both sides by the LCM of 4(x + 3)*

$(x+3)(x-2) = 24$ *Simplify the fractions*

$x^2 + x - 6 - 24 = 0$ *Multiply*

$x^2 + x - 30 = 0$ *Simplify*

$(x+6)(x-5) = 0$ *Factor*

$x + 6 = 0$ or $x - 5 = 0$ *Use zero factor property*

$x = -6$ or $x = 5$ *Solve*

Is it reasonable for x to have the value of –6 in this problem. Why or why not?

Substitute $x = 5$ to find the side lengths of 3 feet and 8 inches.

As in the last problem, after we find the LCM, we multiply both sides of the equation by the LCM. We can solve the resulting equation using algebra or the calculator.

Solve: $\dfrac{7}{x+1} = \dfrac{3}{x^2 - 1}$

What is the LCM?

Now multiply both sides of the equation by the LCM.

$$\cancel{(x+1)}(x-1)\dfrac{7}{\cancel{x+1}} = \cancel{(x+1)(x-1)}\dfrac{3}{\cancel{(x+1)(x-1)}}$$

$$7(x-1) = 3$$

$$7x - 7 = 3$$

$$7x = 10$$

$$x = \dfrac{10}{7}$$

The solution is $x = \dfrac{10}{7}$. Check by substituting $\dfrac{10}{7}$ into the original equation or by solving graphically. On your calculator, graph **Y1** $= 7/(x+1)$ and **Y2** $= 3/(x^2-1)$, and locate the x-value at the point of intersection.

 Solve: $\dfrac{7x}{x^2 + 2x + 1} = \dfrac{3 - 2x}{x^2 - 1}$

What is the LCM?

$$(x+1)^2(x-1)\dfrac{7x}{x^2 + 2x + 1} = (x+1)^2(x-1)\dfrac{3 - 2x}{x^2 - 1}$$

$$7x(x-1) = (x+1)(3-2x)$$

$$7x^2 - 7x = 3x - 2x^2 + 3 - 2x$$

$$9x^2 - 8x - 3 = 0$$

Solve graphically and numerically: $x = $ _____ or $x = $ _____

You Try It

Solve the following equations.

1. $\dfrac{5}{4x^2} = \dfrac{2}{3x}$

2. $\dfrac{7x+2}{x^2+x} = \dfrac{5}{x+1}$

3. Use ratios to calculate the length of the unknown sides in the similar triangles below.

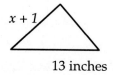
13 inches

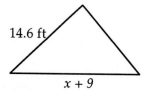
14.6 ft
x + 9

4. Use ratios to calculate the length of the unknown sides in the similar rectangles below.

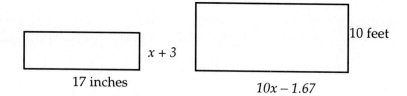

17 inches   x + 3   10 feet
10x − 1.67

Project

For the Birds

CeCe sells bird houses for $24.95 each and sells 1200 a year. Because of increased materials costs, CeCe has decided she must raise her price. Your marketing survey indicated that for every $2.05 increase, she would lose 60 customers. CeCe believes that she must raise her income to at least $31,000, or she must find a full-time job.

1. Let x be the number of $2.05 price increases. Complete the input-output table below.

Number of $2.05 increases	0	1	2	3	4	5	x
Income per bird house	24.95						

2. Write an expression to show the new price per house.

3. Complete the input-output table below:

Number of $2.05 increases	0	1	2	3	4	5	x
Number of Sales	1200						

4. Write an expression to show the decrease in sales.

5. The equation $I = (24.95 + 2.05x)(1200 - 60x)$ gives the total income I according to the number of $2.05 price increases. Let $I = \$31,000$, and solve the equation for x to determine the number of increases needed for an income of $31,000.

6. Prepare an input-output table for your marketing company to help CeCe decide on the best price.

7. What new price would you recommend to CeCe? Justify your recommendation to CeCe.

Summary

During your study of this unit, you:

1. Factored expressions using the distributive law, $ab + ac = a(b + c)$;

2. Factored differences of perfect squares, $a^2 - b^2 = (a - b)(a + b)$;

3. Factored trinomials. A trinomial of the form $ax^2 + bx + c$ is called quadratic. The leading coefficient is a, and the degree is 2. When the lead coefficient is 1, we find factors of the constant, c, whose sum is the middle term, b;

4. Solved quadratic equations of the form $x^2 + bx + c = 0$ by factoring when possible;

5. Solved quadratic equations of the form $ax^2 + bx + c = 0$ numerically and graphically;

6. Used graphical and numerical methods to approximate factors;

7. Added, subtracted, multiplied, and divided rational expressions;

8. Used factoring to solve rational equations; and

9. Solved application problems.

Unit 6 Problems for Practice

A.   Solve algebraically and either graphically or numerically:

1. $\dfrac{x}{7} + \dfrac{2x}{3} = 12$

2.   7 is what percent of 65?

3.   What is 5% of 75?

4.   5 is 300% of what?

5. $\dfrac{\frac{1}{4}}{270} = \dfrac{\frac{1}{5}}{Q}$

6. $\dfrac{12}{25} = \dfrac{p}{100}$

7.   A person riding a Ferris wheel with a diameter of 65 feet travels once around the wheel in 30 seconds. What is the average speed of the person in feet per second?

8.   A college textbook wholesales at $25 and is sold to the student for $36.25. What is the percent of markup?

9.   Similar objects have the same shape, but not necessarily the same size. Similar objects have corresponding parts, and the relationship between the sizes of corresponding parts can be written as a ratio. Each ratio is the same.

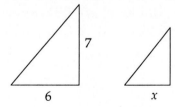

Solve the equation $\dfrac{7}{6} = \dfrac{5}{x}$ to calculate the length of the unknown side.

10.   Use ratios to calculate the length of the unknown side.

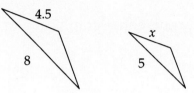

11.   Use ratios to calculate the length of the unknown side.

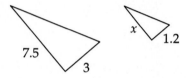

12.   The number of gallons of gas used by a car varies directly with the number of miles driven. If 8 gallons are used to drive 175 miles, how many gallons are needed to drive 148 miles?

13.   The number of revolutions per minute an axle on a vehicle spins varies directly with the speed of the vehicle. If the number of RPMs is 840 at a speed of 60 mph, what is the speed of the vehicle when the RPM is 1149?

14.   Under specific circumstances, the blood alcohol level of a person varies directly with the number of cans of beer consumed. If the person's blood alcohol level is 0.05% after drinking 2 cans of beer, what is the person's blood alcohol level after drinking 5 cans of beer?

B. Solve for the indicated variable:

15. $2x + 3y = 12$ for y

16. $5y - 2x + 20 = 0$ for x

17. $y = kx$ for k

18. $V = \frac{4}{3}\pi r^3$ for π

19. $PV = NkT$ for T

20. $\frac{E_1}{E_2} = \frac{r_2}{r_1}$ for r_1

C. Solve for the indicated variable and complete the I-O table:

21. $10x - 5y = 30$ for y

22. $2x - 5y = 30$ for y

x	y
–5	
–3	
0	
3	
5	

x	y
–10	
–5	
0	
5	
10	

D. Find the slope, y-intercept and x-intercept for each of the following. Sketch a graph for each.

23. $2x + 3y = 6$

24. $5x - 3y = 30$

25. $x = -4$

26. $5y = 20$

E.   Find the equation of each line below.

27. 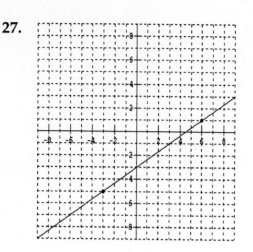   **28.**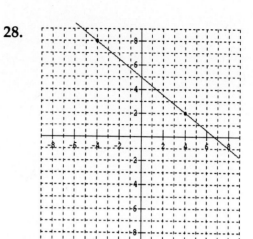

29.   Through the points (0, 2) and (5, 9)

30.   Through the points (2, 7) and (−3, −3)

F.   Factor each of the following:

31. $x^2 + 2x$ **32.** $3x^2 + 3x$

33. $4x^2 y^2 + 8xy$ **34.** $9a^2 b + 6ab^2 - 3ab$

35. $x^2 + 6x + 5$ **36.** $x^2 - x - 6$

37. $x^2 - 2x - 8$ **38.** $x^2 - 6x + 8$

39. $x^2 + 8x - 20$ **40.** $x^2 + 3x + 7$

41. $x^2 - 36$ **42.** $81 - x^2$

43. $4x^2 - 4x - 24$

44. $5x^2 - 180$

45. $7x^2 - 20$

46. $2x^2 - 18$

47. $8 - 2x - x^2$

48. $v^3 + 6v^2 + 5v$

49. $t^3 - t^2 - 6t$

50. $v^2 + 3v + 2$

G. Solve by factoring. Verify graphically:

51. $x^2 = 2x$

52. $3x^2 + 3x = 0$

53. $x^2 = 2x + 8$

54. $x^2 - 6x + 8 = 0$

55. $x^2 + 8x = 20$

56. $x^2 + 3x + 7 = 0$
$7 \; | $
$(x + \quad)(x + \quad) = 0$
prime – No solutions

57. $x^2 - 36 = 0$

58. $81 = x^2$

59. $4x^2 - 4x - 24 = 0$

60. $5x^2 = 80$

61. $v^3 + 6v^2 + 5v = 0$

62. $t^4 - t^3 = 6t^2$

63. The equation $h = -0.005d^2 + 2.67d$, where d is the horizontal distance in feet from the launch point and h is height in feet, approximates the path of a model rocket one windy day. How far from the launch point did the rocket land?

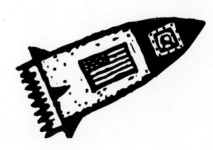

64. The equation $h = -0.64d^2 + 32d$, where d is the horizontal distance in feet from the launch point and h is height in feet, approximates the path of Jeremy's model rocket on a still day. How far from the launch point did the rocket land?

65. Find the length and width of a rectangle with an area of 42 square inches if the width is given as $x - 8$ and the length as $x + 3$.

66. Find the length and width of a rectangle with an area of 40 square inches if the width is given as $x - 4$ and the length as $x + 2$.

67. Find the base and height of a triangle with an area of 24 square feet if the base is given as $x - 3$ and the height as $x + 5$.

68. A baseball is thrown into the air with an initial vertical velocity of 64 feet per second. The equation that gives the height of the baseball above the ground after t seconds is given by the equation $h = 48t - 16t^2$. When will the height be 20 feet above the ground? Let h equal 20 feet and solve graphically.

70-74

H. Solve graphically and/or numerically:

69. $0.3x^2 + 0.33x = 2.346$

70. $8x^2 - 48x + 64 = 0$

Formula

$x = \dfrac{-b \pm \sqrt{b^2 - 4ac}}{2a}$

71. $0.8x^2 - 2.8x + 2 = 3.6$

72. $2x + 4 = x^2$

73. $x^2 = 5 + 2x$

74. $0.5p^2 - 2.5p + 3 = 0$

$\dfrac{-2.04}{.6} = -3.4$

$\dfrac{-.33 + \sqrt{(.33)^2 - 4(.3)(-2.346)}}{2(.3)}$

$\dfrac{.33 + \sqrt{.1089 + 2.8152}}{.6}$

$\dfrac{-.33 + \sqrt{2.924}}{.6}$

$\dfrac{-.33 + 1.71}{.6} = \dfrac{1.38}{.6} = 2.3$

75. If you deposit $1000 in an account with an annual interest rate of r compounded twice a year, then the account balance at the end of the year is given by $A = 1000\left(1 + r + \dfrac{r^2}{4}\right)$. If you want the amount at the end of the year to be $1150, what annual interest rate do you need? Solve graphically or numerically.

76. If $2500 is deposited in an account with an annual interest rate of r compounded twice a year, then the account balance at the end of the year is given by $A = 2500\left(1 + r + \dfrac{r^2}{4}\right)$. If you want the amount at the end of the year to be $2704, what annual interest rate do you need? Solve graphically or numerically.

77. Find the length and width of a rectangle with an area of 48 square inches if the width is given as $x + 2$ and the length as $3x - 4$.

78. Find the base and height of a triangle with an area of 6.5 square feet if the base is given as $x - 3$ and the height as $2x + 5$.

I. Use graphical and numerical methods to approximate the factors of the following:

79. $x^2 + 7x + 12$

80. $225x^2 - 900$

81. $6a^2 + 5a - 6$

82. $4t^2 - 17t + 10$

83. $15u^2 + 19u + 15$

84. $20z^2 - 37z + 15$

J.   Perform the indicated operations.

85. $\dfrac{11}{7a^2} + \dfrac{13}{3a}$

86. $\dfrac{4-3t}{t-t^2} - \dfrac{3}{1-t}$

87. $\dfrac{3}{5t^3} + \dfrac{4}{3t^2}$

88. $\dfrac{7}{r+5} - \dfrac{2r+3}{r^2+5r}$

89. $\dfrac{z^2+2z}{6z} \cdot \dfrac{8z^2}{z+2}$

90. $\dfrac{y+3}{7y^2} \cdot \dfrac{5y}{y^2-9}$

91. $\dfrac{x^2-49}{x^2+3x} \div \dfrac{2x+14}{x^2+2x-3}$

92. $\dfrac{w^2-6w+5}{w^2} \div \dfrac{w^2-25}{w^2+5w}$

93.   The formula $P = \dfrac{1}{A} + \dfrac{1}{B}$ is used by optometrists to calculate how strong to make the lenses for a pair of glasses. Add the terms on the right-hand side of the equation.

94.   Give an alternate formula of the resistance formula $\dfrac{1}{R} = \dfrac{1}{R_1} + \dfrac{1}{R_2}$

by adding the fractions on the right hand side of the equation.

K.   Solve and check:

95. $\dfrac{5}{4c^3} = \dfrac{3}{2c^2}$

96. $\dfrac{3x}{x^2+2x+1} = \dfrac{2-3x}{x^2-1}$

97. $\dfrac{7u-2}{u^2-u} = \dfrac{5}{u-1}$

98. $\dfrac{5}{y-2} + \dfrac{3}{y+2} = \dfrac{y+3}{y^2-4}$

99. In a flight from Portland, Oregon to Cincinnati, Ohio, the plane offered a service for passengers that showed the speed of the wind and the speed of the plane. Mia determined that the plane could fly 1000 km flying with the wind. Flying into the wind, the plane could only travel 800 km in the same amount of time. The speed of the plane in still air is 175 km/hour. Solve the equation $\dfrac{1000}{175+w} = \dfrac{800}{175-w}$ to find the speed of the wind w.

100. Using your calculator, graph $y = \dfrac{1}{x-2}$ on the window $[-4.7, 4.7]_x$ and $[-12, 12]_y$. Trace to determine what value(s) of x make y undefined. Why? Without graphing and tracing, for what value(s) of x will $y = \dfrac{1}{x+5}$ be undefined?

101. According to health reports the ratio of total cholesterol to HDL cholesterol should be less than 4.0. Mike's total cholesterol is 185 mg/dl. If HDL cholesterol is reported to the nearest whole number, what HDL cholesterol should Mike try to reach?

102. Using your calculator, enter $y = \dfrac{1}{(x-2)(x+1)}$ into **Y1**. Numerically, start with a value of $x = -2$ and ΔTbl of 0.1. Determine the x-values that cause the fraction to be undefined.

103. Use ratios to calculate the length of the unknown sides in the similar rectangles below.

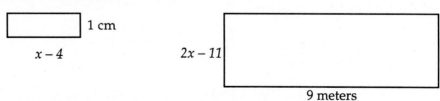

104. Use ratios to calculate the length of the unknown sides in the similar triangles below.

Systems of Linear Equations

7

Upon successful completion of this unit you should be able to:

1. Solve systems of linear equations graphically;

2. Solve systems of linear equations numerically using tables;

3. Solve systems of linear equations algebraically using substitution; and

4. Solve applications using systems of equations.

Introduction

In this unit, we will solve systems of two linear equations in two unknowns. We often find it easier to describe situations using two equations rather than one. We will learn that two lines intersect at one point, at an infinite number of points, or not at all. The point or points where the two lines intersect may be the solution to a real-world problem.

There are a number of methods we can use to solve a system of two linear equations. Solving a system means finding all points (x, y) that satisfy both equations. Graphically, we can locate the point of intersection. Numerically, we can use a table to approximate the solution. Algebraically, we can use the method of substitution.

Graphical and Numerical Solutions

The Delta College Marketing Club sold forty-four tickets to their annual fund raising dinner. Vegetarian dinner tickets sold for $3 each while crab leg dinner tickets were $5 each. If the club raised $158 for their projects, how many of each type ticket did they sell?

Let V equal the number of vegetarian dinner tickets sold. Let C equal the number of crab leg dinner tickets sold.

We can solve the following system to find the answer.

$$V + C = 44$$
$$3V + 5C = 158$$

Justify the preceding equations.

Next, we will solve this system and others graphically and numerically. Later we will find algebraic solutions to systems of equations.

To graph the system, we change each equation into the form $y = mx + b$. We can then graph the lines and trace to find the point of intersection. How do we know the lines will cross?

How many of each type of dinner ticket did the marketing club sell?

Solve for C. Although we could solve for either C or V, we arbitrarily decided to solve each equation for C.

$V + C = 44$ _____

$3V + 5C = 158$ _____

Enter the two equations in **Y1** and **Y2**. Which variable do you enter as x?

Remembering that a maximum of 44 dinners were sold, we select a window that has a maximum of 44 for each variable. To be certain that we can see the x and y axes, we choose –5 to be the minimum for both x and y.

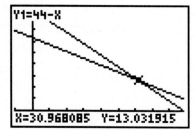

On this graph, the point of intersection is approximately $(x, y) = (30.9, 13)$. We expect the value of V to be between 30 and 31.

We can use a table to improve the estimate. We used 30 for **TblStart** and 0.1 for **ΔTbl** and arrowed down.

X	Y1	Y2
30.7	13.3	13.18
30.8	13.2	13.12
30.9	13.1	13.06
31	13	13
31.1	12.9	12.94
31.2	12.8	12.88
31.3	12.7	12.82

X=30.7

Look at the values of **Y1** and **Y2** in the table. At what value of x are both **Y1** and **Y2** equal?

Both **Y1** and **Y2** were 13 when x was 31. Therefore, the marketing club sold 13 crab leg dinner tickets and 31 vegetarian dinner tickets.

Substitute 31 for V and 13 for C to make sure that these values make *both* equations true.

$V + C = 31 + 13 = 44$, which was the total number of tickets;

$3V + 5C = 3(31) + 5(13) = \$158$, which was the total raised.

Therefore $V = 31$ and $C = 13$ is the solution to the system.

We will now solve some other systems graphically and numerically.

The Upper Valley Community College Computer Club has an imprint machine that produces specialty T-shirts for family reunions, sports teams, and other occasions.

The club sells the T-shirts to various organizations for $7.77 each. Their costs include a setup charge of $135 for each new design plus a blank shirt charge of $2.37 per shirt. How many shirts must the club sell to break even on each new design?

Complete the following table to compare cost and revenue from the sale of various numbers of T-shirts.

Number of T-shirts, n	Cost, C	Revenue, R
1	135 + 2.37(1) = $137.37	7.77(1) = $7.77
2	135 + 2.37( ) =	7.77( ) =
10		
20		
30		
40		
50		
100		

1. How did you calculate the cost of n T-shirts? The revenue from n T-shirts?

2. Which rows of the table have cost greater than revenue?

3. Which rows have cost less than revenue?

4. Which row would first result in a profit?

The **break-even point** occurs when revenue equals cost.

Using the pattern you found in the table, revenue is the money the club earns from sales. The revenue can be expressed as $R(n) = 7.77n$, where n is the number of shirts manufactured. $R(n)$ is a notation showing that R depends upon n. For example, $R(10) = (7.77)(10) = \$77.70$. Cost depends on the setup charge and the T-shirt charge, so we could say $C(n) = 135 + 2.37n$.

To determine the break-even point, solve the equation $C(n) = R(n)$.

$$135 + 2.37n = 7.77n$$

Graphically

To use your calculator to determine the break-even point graphically:

- Enter $C(n)$ in **Y1** and $R(n)$ in **Y2**;

- Adjust the window so that you can see both lines; and

- Trace to find the point of intersection of the lines.

Your graph should be similar to the one at the right, but your intersection point may be slightly different because of your choice of window. The one shown here is $[-7.6, 30]_x$ and $[-25, 225]_y$.

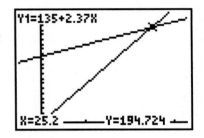

The lines cross at approximately $x = 25.2$ and $y = 194.724$.

This means if the club sells 25.2 T-shirts, they will have revenue and cost of approximately $194.72. Can the club manufacture 25.2 shirts?

To use the calculator table feature to find the break-even point numerically, we can set up a table and find the value of n where **Y1 = Y2**. However, we can also find the value of n where the Revenue – Cost, $R(n) - C(n)$, is nearest to or equal to zero.

Numerically

Use the calculator table feature to solve the system numerically.

- Enter $C(n)$ in **Y1** and $R(n)$ in **Y2**;

- Enter **Y2 – Y1** in **Y3**;

- Press 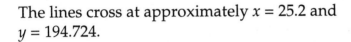, and enter appropriate values for **TblStart** and **ΔTbl**;

- Press 2nd GRAPH to view the table.

Note: **Y1** and **Y2** are accessed by pressing VARS , ▶ for **Y-VARS**, 1 for Function…, and then choosing the desired function.

For this problem, enter 25 for **TblStart** and 0.1 as △**Tbl**;

The first table below shows columns for **Y1** and **Y2**, while the second shows column **Y3**. You can see your results in **Y3** by pressing the right arrow. If you arrow up and down, you will get the following (x, **Y3**) values in the second table: (24, −5.4), (25, 0), and (26, 5.4).

This means the club would have a loss of $5.40 for 24 shirts and a profit of $5.40 for 26 shirts. They have zero profit and zero loss at n = 25 shirts, so they need to manufacture and sell 25 T-shirts to break even.

We find profit by subtracting cost from revenue: $P(n) = R(n) - C(n)$. Notice that **Y3**(n) is the profit for n shirts. Complete the following table:

n	C(n)	R(n)	P(n)
24			
25			
26			
27			
50			
100			

To fill the table, you can press 2nd WINDOW , then press the right arrow to change **Indpnt** to **Ask**. Next press 2nd GRAPH to access the table. Enter your value of x, and the calculator will evaluate **Y1**, **Y2**, and **Y3**.

X	Y₁	Y₂
25	194.25	194.25
25.1	194.49	195.03
25.2	194.72	195.8
25.3	194.96	196.58
25.4	195.2	197.36
25.5	195.44	198.14
25.6	195.67	198.91

X=25

X	Y₃	
25	0	
25.1	.54	
25.2	1.08	
25.3	1.62	
25.4	2.16	
25.5	2.7	
25.6	3.24	

X=25

For this problem, you could have begun with a △**Tbl** of 1, since we don't sell parts of a T-shirt.

Joe vs. Trooper Maura

Joe is speeding south down I-79 away from Pittsburgh. When he is 37 miles from Pittsburgh, Trooper Maura clocks him at 75 mph. Joe goes an additional 0.25 miles when Maura reaches 85 mph for the chase. How long does it take Trooper Maura to catch Joe to issue him a ticket?

Joe's distance: $J(t) = 37.25 + 75t$

Trooper Maura's distance: $M(t) = 37 + 85t$

When Maura catches Joe, the distance they travel will be equal. To determine when Maura catches Joe, first solve the equation:

$$\text{Joe's distance} = \text{Maura's distance}$$
$$J(t) = M(t)$$
$$37.25 + 75t = 37 + 85t$$

Graphically

Enter $J(t)$ in **Y1** and $M(t)$ in **Y2**. Trace to approximate the point of intersection.

The point of intersection appears to be about (0.03, 39.14). This means the trooper would catch Joe after 0.03 hours or 1.8 minutes, approximately 39 miles from Pittsburgh.

Numerically

Enter **Y2 – Y1** in **Y3** and construct a table; begin with $x = 0.02$ and use an increment of 0.001.

At what point does **Y3 = Y2 – Y1** = 0? _____

Complete the following table:

t, hrs	J(t)	M(t)	M(t) – J(t)
0.024			
0.025			
0.026			
0.027			

Which of the table values reflect the fact that Maura passed Joe? Why?

You Try It

1. How many T-shirts does the computer club need to sell to break even if they charge $6.50 per shirt?

2. How many T-shirts does the computer club need to sell to break even if they find a way of reducing the setup charge to $97, and they charge $6.50 per shirt?

3. Where would Trooper Maura catch Joe if she travels at 95 mph?

4. Solve the system: $\begin{aligned} 5x - y &= 1 \\ 4x + 3y &= 16 \end{aligned}$

5. Solve the system: $\begin{aligned} 2x + 3y &= -5 \\ 5x - 2y &= 16 \end{aligned}$

Algebraic Solutions

Although you can often quickly estimate a solution to a system of equations, you may or may not be able to determine the exact answer graphically or numerically. Also, if the slopes of the lines are nearly the same, it is difficult to see the graphical solution. Algebraic methods for solving systems of equations allow us to find exact solutions.

You can solve the equation **Y1 = Y2** to find the solution to a system of equations. We will use algebraic methods to solve systems of equations by solving problems we have already solved graphically and numerically.

Recall that the Delta College Marketing Club sold forty-four tickets for their annual fund raising dinner. Vegetarian dinner tickets sold for $3 each, while crab leg dinner tickets were $5 each. If the club raised $158 for their projects, how many of each type of ticket did they sell?

The two equations are $C = -V + 44$ and $C = \frac{-3}{5}V + \frac{158}{5}$.

We have two different expressions, each equal to C, the number of crab leg dinners. We can set the two expressions equal to each other.

$$C = C$$
$$-V + 44 = \frac{-3}{5}V + \frac{158}{5}$$
$$-5V + 220 = -3V + 158$$
$$62 = 2V$$
$$31 = V$$

If we substitute $V = 31$ in the equation $C = -V + 44$, we have $C = -31 + 44 = 13$.

The solution is $(V, C) = (31, 13)$, which is the same as we found graphically and numerically.

Remember that the Upper Valley Community College Computer Club sells T-shirts to various organizations for $7.77 each. Their costs include a setup charge of $135 for each new design plus a blank shirt charge of $2.37 per shirt. How many shirts must the club sell to break even on each new design?

The equations are

$$y = R(n) = 7.77n$$
$$y = C(n) = 135 + 2.37n$$

We looked for the break-even point, where revenue equals cost. To solve the system, we set $R(n) = C(n)$.

$$R(n) = C(n)$$
$$7.77n = 135 + 2.37n$$
$$5.4n = 135$$
$$n = 25$$

Substitute $n = 25$ in the revenue equation: $R(25) = 7.77(25) = \$194.25$.

Substituting in the cost equation: $C(25) = 135 + 2.37(25) = \$194.25$.

The break-even point occurs at $n = 25$, which is the same solution we found earlier.

Suppose that the computer club sold the T-shirts for $6.50 instead of $7.77. How many shirts do they need to sell to break even?

1. Set up the revenue function: $y = R(n) = $ _____

2. Set up the cost function: $y = C(n) = $ _____

3. Solve $C(n) = R(n)$ algebraically.

The solution to the system is $n \approx 32.7$, but the answer to the problem is 33 T-shirts. Why?

The two equations in the Joe versus Trooper Maura problem are:

$$d = J(t) = 37.25 + 75t$$
$$d = M(t) = 37 + 85t$$

When Maura catches Joe, the distance traveled in the same amount of time will be equal. To determine when Maura catches Joe, first solve the equation:

$$
\begin{aligned}
J(t) &= M(t) \\
37.25 + 75t &= 37 + 85t \\
-10t &= -0.25 \\
t &= 0.025
\end{aligned}
$$

The solution to the equation is $t = 0.025$ hours or 1.5 minutes. The solution to the problem is the distance 39.125 miles south of Pittsburgh. Why?

 Solve the system:

$$
\begin{aligned}
5x - y &= 1 \\
4x + 3y &= 16
\end{aligned}
$$

First, solve each equation for y.

$$
\begin{aligned}
y &= 5x - 1 \\
y &= \frac{-4}{3}x + \frac{16}{3}
\end{aligned}
$$

We want the point (x, y) where the y-value of the first equation equals the y-value of the second equation when the x-values are equal. Set **Y1** = **Y2**.

$$5x - 1 = \frac{-4}{3}x + \frac{16}{3}$$
$$15x - 3 = -4x + 16$$
$$19x = 19$$
$$x = 1$$

You can substitute $x = 1$ into any earlier equation. Substitute $x = 1$ into $y = 5x - 1$, so $y = 5(1) - 1 = 4$. The solution is $(x, y) = (1, 4)$.

More on Substitution

Some equations are not easily solved using the **Y1** = **Y2** method of substitution. Let's consider one.

Solve the system: $\begin{aligned} 2x + 3y &= -5 \\ 5x - 2y &= 16 \end{aligned}$

1.   Solve each equation for y.

2.   Set **Y1** = **Y2**. Solve for x.

3.   Substitute the value of x into either equation, and solve for y.

Next, we will solve one equation for a variable and substitute this expression into the other equation using the same system: $\begin{aligned} 2x + 3y &= -5 \\ 5x - 2y &= 16 \end{aligned}$

Solve one of the equations for one of the unknowns. You have already solved the second equation for y: $y = \frac{5}{2}x - 8$.

Instead of solving both equations for y, we substitute the expression that y equals into the other equation.

$$2x + 3y = -5$$
$$2x + 3\left(\frac{5}{2}x - 8\right) = -5$$

When we substituted $\frac{5}{2}x - 8$ for y in the second equation, the resulting equation has only one unknown. Solve for x:

$$2x + 3\left(\frac{5}{2}x - 8\right) = -5$$
$$2x + \frac{15}{2}x - 24 = -5$$
$$4x + 15x - 48 = -10$$
$$19x \qquad = 38$$
$$x \qquad = 2$$

Substitute $x = 2$ to get the value of y.

The solution is $(x, y) = $ _____ .

Check by substituting these values into *both* original equations:

$$2x + 3y \quad = -5 \qquad\qquad 5x - 2y \quad = 16$$
$$2(2) + 3(-3) \overset{?}{=} -5 \quad \text{and} \quad 5(2) - 2(-3) \overset{?}{=} 16$$
$$-5 \quad = -5 \qquad\qquad\qquad 16 \quad = 16$$

We got the same answer using both methods.

It is sometimes easier to solve for x than to solve for y in some systems.

 Solve the system: $\begin{aligned} 4x + 6y &= 3 \\ x + 3y &= 1 \end{aligned}$

Examine the two equations and decide whether it would be easier to solve the first equation or the second equation for either x or y.

1.   Solve the second equation for x: $x = $ _____

2.   Substitute this expression for x into the first equation to get an equation with only y's.

3. Solve for $y =$ _____. Substitute the y-value into the second equa... solve for x.

4. Check first by substituting the (x, y) values in both equations, then check either graphically or numerically.

You Try It

Solve algebraically and check.

1. Last year, the marketing club raised $180 by selling fifty tickets. They sold vegetarian dinner tickets at $3 each and crab leg dinner tickets at $5 each. How many of each did they sell?

2. The computer club is deciding if they should sell a computerized personality test. The program costs $29.95. To use the college's printer, they must pay $0.25 per test. If they sell each test for $1.50, how many tests does the computer club have to sell to break even?

3. Jon left home heading west at an average of 55 miles per hour. Thirty minutes later, Jeremy left home also heading west at an average of 65 miles per hour. When will Jeremy catch Jon?

4. Solve the system: $\begin{aligned} 3x + 5y &= 1 \\ x + 4y &= -2 \end{aligned}$

5. Solve the system: $\begin{aligned} 5x - 2y &= 16 \\ 4x - y &= 11 \end{aligned}$

...s have unique solutions. Some systems have an infinite number
...ns, and others have no solution at all.

266 ...r Trooper Maura and Joe both traveled at 75 miles an hour, but Joe had a 0.25
mile head start, would Maura catch Joe? If we graph the lines for this situation
we find they are parallel because both lines have a slope of 75 mph. If the two
lines are parallel, we say the system is **inconsistent**. Just as we know that
Maura will never catch Joe, an inconsistent system has no solutions.

 Solve another system:
$$3x + y = 7$$
$$6x + 2y = 10$$

Graphically

Solve each equation for y:
$$y = -3x + 7$$
$$y = -3x + 5$$

Just like in the Maura-and-Joe problem above, the slopes of the lines are the
same. Graph each line. Change the window several times to see if you can find
a point of intersection.

It appears that the lines do not intersect. Lines
with the same slope and different y-intercepts
are parallel. When the lines are parallel, we say
the system is inconsistent and has no solution.
Appearances can be deceiving, so it is important
to solve this problem algebraically to confirm
our suspicion that the lines are parallel.

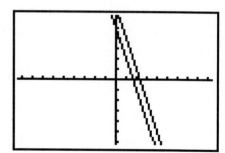

Algebraically

Since we've already solved both equations for y, we can set **Y1 = Y2** and solve.

$$-3x + 7 = -3x + 5$$
$$0 = -2$$

Whenever we solve algebraically and get a false statement (like $0 = -2$), we say
this is an inconsistent system. It has no solution.

Dependent Equations

If the lines in a system of equations have the same graph or coincide, we say that the pair of equations is dependent. A system with **dependent equations** has an infinite number of solutions.

 Solve the system: $\begin{aligned} 3x + y &= 7 \\ 6x + 2y &= 14 \end{aligned}$

Graphically

Solve each equation for y:

$$y = -3x + 7$$
$$y = -3x + 7$$

When solved for y, it is obvious that the equations represent the same line. When two lines have the same slope and y-intercept, we can also say the lines coincide. We say that the equations are dependent, and the system has an infinite number of solutions. In fact, every point on the line is a solution.

Sketch the graph and complete the following table to find some of the solutions to the system.

x	−3	−1	0	4	7
y					

Algebraically

We've already solved the first equation for y: $y = 7 - 3x$.

Substitute the expression $7 - 3x$ in the second equation for the y and solve for x:

$$\begin{aligned} 6x + 2(7 - 3x) &= 14 \\ 6x + 14 - 6x &= 14 \\ 14 &= 14 \end{aligned}$$

Whenever we solve algebraically and get a true statement (like 14 = 14), for whatever value of x we choose, we say that the equations are dependent. The system has an infinite number of solutions.

You Try It

Solve the following systems graphically and algebraically.

1.
$$5x + y = 7$$
$$15x + 3y = 20$$

2.
$$5x + y = 6$$
$$3x + 3y = 6$$

3.
$$3x + 3y = 6$$
$$x + y = 2$$

Project

Apple Carpet Company

Memorandum #7   **Your Town, U.S.A.**

TO: Team Members

FROM: Your Supervisor

RE: Cellular-Phone Company

Mid-management has met with representatives of two different cellular-phone companies. Company A will charge us $14.95 per month plus 35 cents per minute for each phone. Company B will charge us $35.95 a month for 30 minutes, with 25 cents for each minute over 30 minutes for each phone.

How many minutes would we have to use to prefer Company B over Company A? The boss wants you to write a report supporting your conclusions and to present all the data.

1.   Complete the following table:

Minutes of Use	Company A Fee	Company B Fee
10		
20		
30		
40		
50		
n		

2.   What equation represents the first thirty minutes of usage from Company B?

3.   Write an equation that represents cost as a function of time over 30 minutes for each company.

4.   Graph the equations.

Summary

During your study of this unit, you have:

1. Solved a system of two linear equations in two unknowns graphically by locating the point of intersection;

2. Solved these systems numerically using tables by finding the x-value where the y-values were equal;

3. Solved algebraically by substitution;

4. Learned that inconsistent systems have no solution. Inconsistent systems have two parallel lines with the same slope and different y-intercepts.

5. Learned that a system with dependent equations has an infinite number of solutions. The lines have the same slope and y-intercept. The equations represent the same line (the lines coincide).

6. Solved application problems using systems of equations.

Unit 7 Problems for Practice

A.   Review Problems. Follow the instructions.

1.   Calculate the area of a circle with radius of 5 centimeters.

2.   Complete the following patterns:

a) $\dfrac{2}{3}, \dfrac{5}{3}, \dfrac{8}{3}$, _____ , _____ , _____

b)   1, 4, 9, 16, _____ , _____ , _____

3.   Complete the input-output table and plot the points. Are the points linear?

x	$y = -2x - 5$
-2	
-1	
0	
1	
2	

4.   Solve $2(-2x + 3) = x - 5$ algebraically. Verify graphically or numerically.

5.   Calculate the area and perimeter of this right triangle.

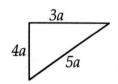

6.   Write this in scientific notation and then simplify:

$$\dfrac{(0.000003)(7,000,000)}{100,000}$$

7.   Five is what percent of 25?

8.   Solve: $\dfrac{25}{75} = \dfrac{5}{L}$

9.   Solve by factoring: $p^2 - 2p - 8 = 0$

10.   Approximate the factors by graphing: $10x^2 - 5x - 7$

B.   Review Problems. Follow the instructions.

11.   Complete an input-output table to solve $75 - 11b = 113.5$.

12.   A local factory job pays $290 a week for summer employment with a $0.75 bonus per unit assembled. If you want to earn $320 before taxes, write the equation that represents your earnings and solve it to determine how many units you need to assemble.

13.   Solve $5 + 3x = 7 - 2x$ algebraically.

14.   Simplify using only positive exponents: $(-3c^5 d^0)(-5c^{-7} d^3)$

15.   Pluto's orbit around the sun is 5,900,000,000 kilometers. How far does it travel in 25 orbits? Write this in scientific notation.

16.   Solve: $\dfrac{7}{15} + \dfrac{x}{3} = \dfrac{5}{2} - \dfrac{x}{4}$

17.   Solve by factoring: $v^2 - 4v - 12 = 9$

18.   The number of gallons of gas used by a car varies directly with the number of miles driven. If 6.8 gallons are used to drive 175 miles, how many gallons are needed to drive 148 miles?

19.   Graph the line through (–2, 5) and (3, 8). Calculate the slope using the triangle method.

20.   Graph the line with a slope of –2 that passes through (–5, 3).

C. Solve the following linear systems graphically and numerically.

21. $y = x + 3$

$y = -2x + 12$

22. $y = \frac{2}{3}x + 3$

$y = \frac{-1}{3}x + 4$

23. $x + 2y = 5$

$3x - y = 8$

24. $2x + y = -1$

$x + 2y = 4$

25. $2x + y = 2$

$3x - y = 8$

26. $x + y = 7$

$5x - 3y = -5$

27. $-2x + y = 6$

$x + 2y = 7$

28. $3x - y = 7$

$x - 2y = 9$

29. $2x + 3y = 5$

$4x - 6y = 0$

30. $5x - 3y = 8$

$10x + 9y = -2$

31. $x + y = 3$

$2x + 2y = 10$

32. $2x - 3y = 9$

$4x - 6y = 12$

33. $x + y = 3$

$3x + 3y = 9$

34. $5x - y = 4$

$10x - 2y = 8$

35. The Mountain Community College Non-Traditional Student Club sold pies to raise funds. Pecan pies sold for $5 each while fruit pies sold for $4 each. If the club raised $197 from the sale of 45 pies, how many of each kind did they sell?

36. Jana leaves Chicago heading southwest averaging 50 mph. An hour later, Henry follows averaging 55 mph. When will Henry catch Jana?

37. The Beta Club sells caps to other student organizations. The setup charge for each new cap design is $30, and blank caps cost the club $2.77 each. If the club sells the caps for $5.99 each, how many do they need to sell to break even on each new design?

38. Tiny Tot's manufactures deluxe and economy tricycles. Deluxe tricycles take 3 hours in shop 1 and 5 hours in shop 2. Economy tricycles take 2 hours in shop 1 and 3 hours in shop 2. Each week there are 120 hours available in shop 1 and 190 hours available in shop 2. How many of each type tricycle can the company produce per week?

39. Bart is trying to decide between pager services. Company A offers an activation fee of $89.95 a year with 400 free pages a month. Company B offers an activation fee of $49.95 with unlimited pages at $0.20 a page. Bart wants you to tell him how many pages he could receive a month before Company A's plan is better.

40. Ben's retirement contribution of $7200 can go into TIAA or CREF. TIAA is safer, but only pays 8.6% interest. CREF is riskier, but averages interest of 10.5%. He estimates that he needs to earn $700 in interest to meet his goals. How should he allocate his $7200?

D. Solve the following linear systems algebraically.

41. $x + 2y = 4$
    $5x - y = 8$

42. $2x - y = -1$
    $x + 2y = 8$

43.
$$y = x + 3$$
$$y = -2x + 12$$

44.
$$3y = 2x + 9$$
$$3y = -1x + 12$$

45.
$$2x + y = 2$$
$$x + 2y = 7$$

46.
$$x + y = 6$$
$$5x - 3y = 8$$

47.
$$2x + 3y = 5$$
$$4x - 6y = -62$$

48.
$$5x - 3y = 8$$
$$10x + 9y = -2$$

49.
$$y = 3 - x$$
$$2x + 2y = 5$$

50.
$$x + y = 3$$
$$3x = 6 - 3y$$

51.
$$2x - 3y = 9$$
$$4x - 6y = 12$$

52.
$$5x - y = 4$$
$$10x - 2y = 8$$

53.   The Valley Community College Non-Traditional Student Club sold baked goods to raise funds. Cookies sold for $0.25 each, while brownies sold for $0.60 each. If the club raised $47.25 from the sale of 105 baked goods, how many of each kind did they sell?

54.   Julie leaves Huntsville on the interstate heading south averaging 72 mph. A quarter of an hour later, Art follows averaging 79 mph. When will Art catch Julie?

55.   The Beta Club sells tote bags to other student organizations. The setup charge for each new design is $65 and blank bags cost the club $6.37 each. If the club sells the bags for $12.50 each, how many do they need to sell to break even on each new design?

56. Lakeland Homes manufactures deluxe and economy homes. Deluxe homes take 5 man-hours in station 1 and 8 man-hours in station 2. Economy homes take 5 man-hours in station 1 and 6 man-hours in station 2. Each week there are 280 man-hours available in station 1 and 400 man-hours available in station 2. How many of each type home can the company produce per week?

57. Tim's cellular-phone provider has offered him his choice of two new plans. The first plan requires a monthly fee of $19.95 plus $0.20 a minute for connection time. The second plan requires no minimum fee, but charges $0.50 a minute. At how many minutes of connection time are the prices of the two plans equal?

Radicals and Radical Equations

Upon successful completion of this unit, you should be able to:

1. Evaluate radical expressions;

2. Simplify radical expressions;

3. Add, subtract, multiply, and divide radical expressions;

4. Connect fractional exponents with radicals;

5. Solve radical equations graphically, numerically, and algebraically; and

6. Solve applications.

Introduction

Radicals occur in many problems. These examples illustrate applications where you might need to work with radicals.

- The escape velocity for a spacecraft is $v = \sqrt{2gR}$, where g is the planet's acceleration due to gravity and R is the radius of the planet.

- The hypotenuse of a right triangle with sides x and y is $H = \sqrt{x^2 + y^2}$.

- The period of a simple pendulum is $T = 2\pi\sqrt{\dfrac{L}{g}}$, where L is the length of the pendulum and g is the acceleration due to gravity where the pendulum is located.

In this unit, we will work problems with these and with other applications of radicals. We will also solve equations with radicals.

Simplifying Radicals

Suppose we have a right triangle with sides x and y. We stated above that the hypotenuse of a right triangle is given by $H = \sqrt{x^2 + y^2}$. The hypotenuse is the longest side of a right triangle. It is also the side opposite the right angle. Consider the following table.

x	y	H
1	2	$\sqrt{5}$
1	3	$\sqrt{10}$
3	3	$\sqrt{18}$
3	4	$\sqrt{25}$
7	8	$\sqrt{113}$
5	12	$\sqrt{169}$

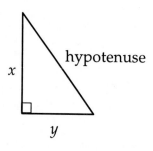

Some of the values of H in the table can be simplified while others cannot. Have you looked at textbooks and noticed numbers like $\sqrt{2}$, $\sqrt{3}$, $\sqrt{5}$, and $\sqrt{7}$? Where do such numbers come from, and why aren't they expressed as fractions or decimals?

By now you have realized that there are many more numbers without integer square roots than with them. However, many numbers have factors that are perfect squares. We can simplify these numbers while maintaining the exact value.

What is the hypotenuse of a right triangle with sides of length 2 cm and 4 cm? Recall that the hypotenuse of a right triangle with sides x and y is $H = \sqrt{x^2 + y^2}$.

$$H = \sqrt{2^2 + 4^2} = \sqrt{4 + 16} = \sqrt{20}$$

To simplify $\sqrt{20}$, we first need to find any factors of 20 that are perfect squares. We could write 20 as any of the following: $1 \cdot 20 = 2 \cdot 10 = 4 \cdot 5$. Of these factors, 4 is the only perfect square. We can rewrite $\sqrt{20} = \sqrt{4 \cdot 5} = \sqrt{4} \sqrt{5}$. Taking the square root of 4, we have $\sqrt{20} = 2\sqrt{5}$. Evaluate both of these expressions in your calculator, and compare the results.

> Let a and b be greater than or equal to zero. Then
>
> $$\sqrt{ab} = \sqrt{a} \sqrt{b}$$

1.   What is the hypotenuse of a right triangle with sides of 3 inches and 5 inches?

$$H = \sqrt{3^2 + 5^2} = \sqrt{9 + 25} = \sqrt{34}$$

The integer $34 = 1 \cdot 34 = 2 \cdot 17$. Since none of these factors is a perfect square, $\sqrt{34}$ cannot be simplified further.

2.   What is the hypotenuse of a right triangle with sides of 4 m and 8 m?

$$H = \sqrt{4^2 + 8^2} = \sqrt{16 + 64} = \sqrt{80}$$

Since 80 has many factors, it might be easier to examine our list of perfect squares to decide if any are factors of 80. From the list of perfect squares smaller than 80, we have 1, 4, 9, 16, 25,

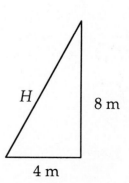

H

8 m

4 m

36, 49, and 64. Of these perfect squares, 80 is divisible by 1, 4, and 16. We select 16, because it is the largest perfect square factor, and $80 \div 16 = 5$. We can write $\sqrt{80} = \sqrt{16 \cdot 5} = \sqrt{16}\sqrt{5} = 4\sqrt{5}$. The exact length of a hypotenuse of a right triangle with sides of 4 m and 8 m is $4\sqrt{5}$ m.

 Simplify the following:

1. $\sqrt{8} = \sqrt{4 \cdot 2} = \sqrt{4}\sqrt{2} = 2\sqrt{2}$

   Eight has a factor of 4, a perfect square. The square root $\sqrt{8}$ simplifies to $2\sqrt{2}$.

2. $\sqrt{6} = \sqrt{3}\sqrt{2}$

   Neither factor of 6 is a perfect square. $\sqrt{6}$ is in simplest form.

3. $\sqrt{18} = \sqrt{9 \cdot 2} = \sqrt{9}\sqrt{2} = 3\sqrt{2}$

   The factor 9 has a square root of 3. $3\sqrt{2}$ is in simplest form.

> A square root is considered simplified if its radicand contains no perfect square factors (other than 1).

You Try It

Write each radicand with a perfect square factor, if possible. Simplify.

1. $\sqrt{20}$ 2. $\sqrt{15}$ 3. $\sqrt{32}$

4. $\sqrt{40}$ 5. $\sqrt{200}$

6. Find the exact length of the hypotenuse of a right triangle with sides 4 inches and 6 inches in length.

Distance Between Two Points

In some applications we need to find the distance between two points plotted on a rectangular coordinate system. We can do this by finding the hypotenuse of a right triangle.

Find the distance between the points (–1, 1) and (4, 3).

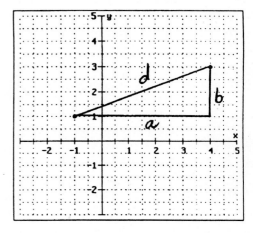

First, we plotted the points. We then drew a right triangle with the line segment connecting (–1, 1) and (4, 3) as the hypotenuse. This is one of the two possible right triangles. Can you find the other one?

Notice that the line segment between the points is labeled *d* for distance. Side *a* is parallel to the *x*-axis, and side *b* is parallel to the *y*-axis. Label each endpoint with its coordinates.

Find the length of side *a* by counting the units and also by subtracting the *x*-coordinates. *a* = _____

Find the length of side *b* by counting the units and also by subtracting the *y*-coordinates. *b* = _____

We now have the lengths of two sides of a right triangle. Express the hypotenuse in simplest form and then as a decimal approximation.

$$d = \sqrt{a^2 + b^2} \ = \ \underline{\hspace{2cm}} \ \approx \ \underline{\hspace{2cm}}$$

You Try It

Find the distance between the following points.

1. (2, 1) and (4, 5)

2. (–3, –3) and (5, 7)

3. (4, 8) and (–3, 8)

4. (–5, 7) and (–5, –5)

5. (–7, 7) and (7, –7)

Operations with Square Roots

Sometimes we need to simplify radical expressions involving addition, subtraction, multiplication, and division. After removing perfect square factors from each individual radical, we add or subtract radicals by combining like terms. Like term radicals have exactly the same radicand and index. We simplify radicals before we add or subtract like terms.

> If c is greater than or equal to zero, then
>
> $$a\sqrt{c} + b\sqrt{c} = (a+b)\sqrt{c}$$

 Combine like terms to simplify.

1. $3\sqrt{2} + 5\sqrt{2} = 8\sqrt{2}$

    We added the integer factors of the terms with $\sqrt{2}$ to get $3 + 5 = 8$.

2. $3\sqrt{2} + 4\sqrt{5} + 5\sqrt{2} + 7\sqrt{5} = \underline{\hspace{1cm}}\sqrt{2} + \underline{\hspace{1cm}}\sqrt{5}$

3. $3\sqrt{12} + 4\sqrt{27} = 3\sqrt{4}\sqrt{3} + 4\sqrt{9}\sqrt{3} = \underline{\hspace{1cm}}\sqrt{3} + \underline{\hspace{1cm}}\sqrt{3} =$

    We first simplified $\sqrt{12}$ and $\sqrt{27}$. This left us with like terms of $\sqrt{3}$.

4. $3\sqrt{12} + 5\sqrt{8} - 4\sqrt{27} - 7\sqrt{50} =$

To multiply expressions containing radicals, we multiply the non-radical factors and then the radical factors together.

> If b and d are greater than or equal to zero, then
>
> $$a\sqrt{b} \cdot c\sqrt{d} = ac\sqrt{bd}$$

Multiply each of the following and express the answer in simplest form.

1. $5\sqrt{8} \cdot 3\sqrt{2} = 15\sqrt{16} = 15 \cdot 4 = 60$

   $(3)(5) = 15$. The radicands $(8)(2) = 16$. The expression $15\sqrt{16}$ simplifies to $(15)(4) = 60$.

2. $5\sqrt{8}\left(3\sqrt{2} + 5\sqrt{8}\right) = 5\sqrt{8} \cdot 3\sqrt{2} + 5\sqrt{8} \cdot 5\sqrt{8}$ *Distributive law*

   $\qquad\qquad\qquad = 15\sqrt{16} + 25\sqrt{64}$ *Multiply radicals*

   $\qquad\qquad\qquad = 15 \cdot 4 + 25 \cdot 8$ *Simplify*

   $\qquad\qquad\qquad = 260$ *Multiply and add*

3. $\left(3 + 5\sqrt{2}\right)\left(5 - 4\sqrt{8}\right) = 3 \cdot 5 - 3 \cdot 4\sqrt{8} + 5\sqrt{2} \cdot 5 - 5\sqrt{2} \cdot 4\sqrt{8}$

   $\qquad\qquad\qquad\qquad =$

   The last result $-65 + \sqrt{2}$ is the product in simplest form. It is the exact value of the product. Use your calculator to approximate its value. Now, approximate the product using your calculator, and compare your results.

You Try It

Simplify the following radical operations.

1. $3\sqrt{20} + 4\sqrt{27} - 3\sqrt{12} + 5\sqrt{8}$

2. $\left(5\sqrt{18}\right)\left(3\sqrt{2}\right)$

3. $\left(5\sqrt{6}\right)\left(3\sqrt{2}\right)$

4. $5\sqrt{6}\left(3\sqrt{2} + 5\sqrt{8}\right)$

5. $\left(3\sqrt{8} + 5\sqrt{2}\right)\left(5\sqrt{3} - 4\sqrt{8}\right)$

6. $\left(\sqrt{2} + \sqrt{3}\right)\left(\sqrt{2} - \sqrt{3}\right)$

Division of Radicals

Division of radicals is similar to multiplication. We divide the non-radical factors and divide the radicands of any radicals with the same index. However, sometimes the quotient is not an integer.

> If $a \geq 0$ and $b > 0$, then
>
> $$\frac{\sqrt{a}}{\sqrt{b}} = \sqrt{\frac{a}{b}}$$

 Express your answers in simplest form.

1. $\dfrac{15\sqrt{8}}{3\sqrt{2}} = \left(\dfrac{15}{3}\right)\left(\dfrac{\sqrt{8}}{\sqrt{2}}\right) = 5\left(\sqrt{\dfrac{8}{2}}\right) = 5\sqrt{4} = 5(2) = 10$

    The radicand $\dfrac{8}{2} = 4$.

2. $\dfrac{5\sqrt{18}}{3\sqrt{6}} =$

3. Give a reason for each step to the right of the work.

    $\dfrac{14\sqrt{15} - 21\sqrt{5}}{7\sqrt{3}} = \dfrac{14\sqrt{15}}{7\sqrt{3}} - \dfrac{21\sqrt{5}}{7\sqrt{3}}$

    $= \left(\dfrac{14}{7}\right)\left(\dfrac{\sqrt{15}}{\sqrt{3}}\right) - \left(\dfrac{21}{7}\right)\left(\dfrac{\sqrt{5}}{\sqrt{3}}\right)$

    $= 2\sqrt{\dfrac{15}{3}} - 3\sqrt{\dfrac{5}{3}}$

    $= 2\sqrt{5} - 3\sqrt{\dfrac{5}{3}}$

In **3**, the radicands had no common factors in the second term, and the quotient was not an integer. Therefore, it could not be simplified. Simplified radicals do not contain fractions or have radicals in the denominator. To simplify these expressions, we use a method called **rationalization**. To rationalize the denominator with a square root, we multiply both numerator and denominator of the expression to create a perfect square radicand in the denominator.

1. Simplify: $\dfrac{3}{\sqrt{5}}$

a) What is $\sqrt{5} \cdot \sqrt{5}$?

b) If we want to multiply the denominator by $\sqrt{5}$, by what number must we multiply the numerator to maintain the value of the fraction?

$$\frac{3}{\sqrt{5}} = \left(\frac{3}{\sqrt{5}}\right)\left(\frac{\sqrt{5}}{\sqrt{5}}\right) = \frac{3\sqrt{5}}{\sqrt{25}} = \frac{3\sqrt{5}}{5}$$

This fraction is considered to be in *simplest* form, because we were able to eliminate the radical in the denominator. The elimination of any radicals in the denominator improves the accuracy of any approximation, since you are not dividing by an approximation. It was also very difficult to divide by a good decimal approximation of square roots before we had calculators. Now, in many applications, approximations will be calculated without rationalizing the denominator.

2. Simplify: $\dfrac{3\sqrt{7}}{\sqrt{10}} =$

Multiply numerator and denominator by $\sqrt{10}$, and simplify.

3. Simplify: $\dfrac{3\sqrt{5} + 2\sqrt{3}}{\sqrt{6}} =$

You Try It

Simplify the following:

1. $\dfrac{3\sqrt{50}}{\sqrt{10}}$

2. $\dfrac{3\sqrt{6}}{\sqrt{2}}$

3. $\dfrac{8\sqrt{6} + 4\sqrt{12}}{2\sqrt{3}}$

4. $\dfrac{8\sqrt{6} + 5\sqrt{12}}{2\sqrt{8}}$

5. $\dfrac{3}{\sqrt{6}}$

6. $\dfrac{3 - 2\sqrt{6}}{\sqrt{5}}$

Radicals and Fractional Exponents

Radicals can be written using radical signs or with fractional exponents. We evaluate radicals on the calculator using either the [MATH] menu or using the [^] key.

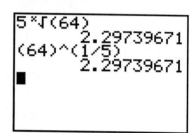

To evaluate $\sqrt[5]{64}$ using the [MATH] menu, first enter 5 on the home screen. Then press [MATH], and arrow down so 5 is backlit, and press [ENTER] or [5]. Close the parentheses, and press [ENTER].

To evaluate $(64)^{\frac{1}{5}}$ using the [^] key, enter the radicand and then press [^] and enter the (1/5) as shown. Note that the answers are exactly the same to eight decimal places.

Use your calculator to evaluate these.

1. a) $(36)^{\frac{1}{2}}$ b) $\sqrt{36}$

2. a) $(27)^{\frac{1}{3}}$ b) $\sqrt[3]{27}$

3. a) $(1024)^{\frac{1}{5}}$ c) $\sqrt[5]{1024}$

Use the results from each of the above exercises to answer the questions below.

4. What fractional exponent did you use to evaluate square roots?

5. What fractional exponent did you use to evaluate cube roots?

6. What fractional exponent did you use to evaluate fifth roots?

7. What fractional exponent would you use to evaluate the *n*th root?

Finding the hypotenuse of a right triangle is an application we solved using a square root or an exponent of $\frac{1}{2}$. Next, we will explore applications that require other roots.

A sphere has a volume of 36π cm³. What is its radius?

$$V = \frac{4}{3}\pi r^3 \qquad \textit{Volume of a sphere formula}$$

$$36\pi = \frac{4}{3}\pi r^3 \qquad \textit{Substitute the given volume for V}$$

$$\frac{3}{4\pi}(36\pi) = r^3 \qquad \textit{Solve for } r^3$$

$$27 = r^3 \qquad \textit{Simplify}$$

$$r^3 = 27$$

To find the radius, we take the cube root of both sides:

$$(r^3)^{\frac{1}{3}} = (27)^{\frac{1}{3}} \quad \text{or} \quad \sqrt[3]{r^3} = \sqrt[3]{27}$$
$$r = 3 \text{ cm}$$

You can evaluate this on your calculator using the $\boxed{\text{MATH}}$ menu or using the caret (power) key and the exponent of $1 \div 3$, which must be enclosed in parentheses. What would happen if you forgot the parentheses?

An investment of $5000 earned $825 in interest when compounded yearly for five years. What was the rate of interest?

Let r be the rate of interest expressed as a decimal. After 5 years, the amount in the account is $5000 + $825 = $5825.

$A = P(1 + r)^t$, where A is the amount in the account, P is the principal (the beginning investment), r is the rate as a decimal, and t is the number of years for money compounded yearly.

Give a reason for each step to the right of the work.

$$5825 = 5000(1 + r)^5$$

$$\frac{5825}{5000} = (1 + r)^5$$

$$\sqrt[5]{\frac{5825}{5000}} = \sqrt[5]{(1 + r)^5}$$

$$\sqrt[5]{\frac{5825}{5000}} = 1 + r$$

$$\sqrt[5]{\frac{5825}{5000}} - 1 = r$$

$$r \approx 0.0310 = 3.1\% \text{ interest}$$

You Try It

Evaluate the following using roots and fractional exponents, rounding to two decimal places:

1. $(125)^{\frac{1}{2}} =$

2. $\sqrt[5]{48} =$

3. $(48)^{\frac{1}{3}} =$

4. Determine the radius of a sphere with a volume of 600π cubic inches.

5. Calculate the rate of interest of a $10,000 investment that earns $4000 interest when compounded yearly for four years.

The period of a simple pendulum (similar to the pendulum in a grandfather clock) is $T = 2\pi\sqrt{\dfrac{L}{g}}$, where L is the length of the pendulum in *meters* and g is the acceleration due to gravity. The period is the length of time it takes the pendulum to return to the same point. What is the period of a clock pendulum that is 30 centimeters long? Use $g = 9.8$ m/s².

$$T = 2\pi\sqrt{\frac{L}{g}}$$

$$T = 2\pi\sqrt{\frac{0.3\,\text{meters}}{9.8\,\text{meters}/\,\text{sec}^2}}$$

$$T \approx 1.1 \ \text{seconds}$$

You Try It

1. Find the area of a triangle with sides of 4, 5, and 6 feet.

2. Find the velocity of a satellite that is orbiting 30,000 miles from the center of the earth.

3. Find the third side of a right triangle if the hypotenuse is 8 cm and one side is 4 cm.

4. Find the period of a clock with a pendulum that is 36 inches long (1 meter = 39.37 inches).

Radical Equations

Expressions with radicals often occur in equations. These can be solved graphically, numerically, and algebraically. Algebraically, to solve an equation with a square root, we can square both sides of the equation. We cube each side to solve cube root equations. To solve nth root equations, we raise both sides to the nth power.

Suppose a 24-foot ladder is leaning against a house. The manufacturer recommends a limit of how far away from the building the foot of the ladder may be placed. If the foot of the ladder is 12 feet from the ground, how far up the house is the top of the ladder? What if the foot of the ladder is 15 feet from the building?

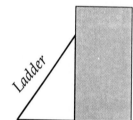

The Pythagorean theorem states that $x^2 + y^2 = r^2$, where x and y are the sides and r is the hypotenuse of a right triangle. Let y be the distance from the foot of the ladder to the house and x be the distance from the top of the ladder to the ground. We want to find the value x when y is 12 and again when y is 15.

$$
\begin{aligned}
x^2 + y^2 &= r^2 && \textit{Pythagorean theorem} \\
x^2 + y^2 &= 24^2 && \textit{Substitute 24 for r} \\
y^2 &= 576 - x^2 && \textit{Solve for } y^2 \\
y &= \pm\sqrt{576 - x^2} && \textit{Solve for r} \\
y &= \sqrt{576 - x^2} && \textit{Only the positive value is reasonable}
\end{aligned}
$$

Why are only positive values reasonable for this problem?

Graphically

Enter $\sqrt{576 - x^2}$ in **Y1**, and trace to estimate the values of x for the given values of y.

For this problem, you might set minimum values for x and y to –1 so you can see the axes and the maximum values to 24 since the ladder can reach no farther than 24 feet up the wall or away from the wall. Press GRAPH, and look at the graph. If you are having difficulties setting the window, you might try changing only one value of the window at a time. If you are not satisfied with the graph,

now change **Xmax**. Finally, adjust the **Xmin** and **Ymin** until your graph shows the values you want.

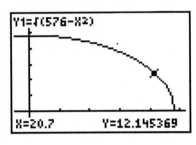   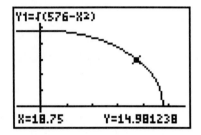

$x \approx$ _____ when $y \approx 12$ $x \approx$ _____ when $y \approx 15$

Record your window: [,   ]$_x$ and [,   ]$_y$

Numerically

To obtain a closer approximation, you can use the table feature of your calculator to determine the value of x when y is 12.

Enter 20.7 in **TblStart** and 0.01 as Δ**Tbl**.

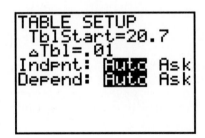

At what point is the value of y closest to 12? $(x, y) =$ _____

To determine the x-value, where y is closest to 15, enter _____ for **TblStart** and 0.01 as Δ**Tbl**.

When is the value of y closest to 15? $(x, y) =$ _____

Algebraically

For $y = 12$: Give a reason for each step to the right of the work.

$$\sqrt{576 - x^2} = 12$$
$$576 - x^2 = 144$$
$$x^2 = 576 - 144$$
$$x = (576 - 144)^{\frac{1}{2}}$$
$$x \approx 20.78 \text{ feet}$$

For $y = 15$: Work each step and describe what you did.

$$\sqrt{576 - x^2} = 15$$

Orbiting the Earth

We learned earlier that the velocity of a satellite orbiting R meters from the center of the earth is given by $v = \sqrt{\dfrac{4.0 \times 10^{14}}{R}}$ m/sec. How far is the satellite from the center of the earth if its velocity is given by 3×10^3 meters per second?

We want to find the distance between the satellite and the center of the earth when $v = 3 \times 10^3$ m/sec.

$$v = \sqrt{\frac{4.0 \times 10^{14}}{R}}$$

$$3 \times 10^3 = \sqrt{\frac{4.0 \times 10^{14}}{R}}$$

$$(3 \times 10^3)^2 = \left(\sqrt{\frac{4.0 \times 10^{14}}{R}} \right)^2$$

$$9 \times 10^6 = \frac{4.0 \times 10^{14}}{R}$$

$$R = \frac{4.0 \times 10^{14}}{9 \times 10^6}$$

$$R \approx 4.4 \times 10^7 \text{ meters}$$ *Now, verify graphically or numerically.*

Find the value of R in the velocity formula when $v = 2 \times 10^4$ m/sec.

 Solve $\sqrt[3]{2x+3} = 3$ algebraically and verify graphically.

Algebraically

$$\left(\sqrt[3]{2x+3}\right)^3 = 3^3 \qquad \text{Cube both sides}$$
$$2x+3 = 27 \qquad \text{Simplify}$$
$$2x = 24$$
$$x = 12$$

Graphically

Enter $\sqrt[3]{2x+3}$ in **Y1** and 3 in **Y2**.

Where do they intersect?

Check: $\sqrt{2x+3} = \sqrt{2(12)+3} = \sqrt{27} = 3$

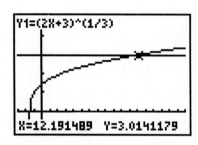

You Try It

Solve algebraically and verify graphically or numerically.

1. $\sqrt{2x-3} = 4$

2. $\sqrt{x^2 - 3} = 6$

3. $\sqrt[4]{2x+5} = 6$

4. $\sqrt[3]{3-x^2} = 7$

5. How far from the center of the earth is a satellite orbiting at a velocity of 5×10^2 meters per second?

6. What is the length of a clock's pendulum if the period is 2 seconds? Use $g = 9.8$ m/s^2.

Project

Neighborhood Park Association

Memorandum #8   **Your Town, U.S.A.**

TO: Association Members

FROM: Chairperson

RE: Area of the Park

In preparing our documents, it came to my attention that we do not know the dimensions or area of the park. This is an important omission that needs to be remedied immediately. We have a scale drawing of the park with the scale given. Please follow the instructions below and provide me with a final report.

1. Draw a right triangle with the line segment from (1, 3) to (3, 1) as its hypotenuse. Use this triangle to find the length of side *a*.

2. Draw a right triangle with the line segment (3, 1) to (5, 5) as its hypotenuse. Use this triangle to find the length of side *b*.

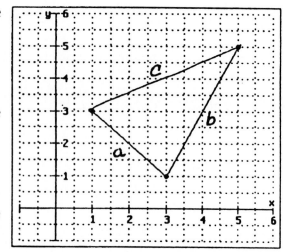

3. Draw a right triangle with the line segment (1, 3) to (5, 5) as its hypotenuse. Use this triangle to find the length of side *c*.

4. Use Heron's formula to find the area of the drawing.

5. The scale on the drawing is 5 units to 30 meters. Change the length of each side of the drawing into the length of the sides of the park.

   a = _____ *b* = _____ *c* = _____

6. Use Heron's formula to find the area of the park.

Summary

During your study of this unit, you have:

1. Evaluated radical expressions;

2. Simplified square roots. A simplified square root expression has

    a) no perfect square factors in the radicand;
    b) no fractions in the radicand;
    c) no like terms; and
    d) no square roots in the denominator.

3. Added and subtracted radical expressions. If $c \geq 0$, then
    $a\sqrt{c} + b\sqrt{c} = (a+b)\sqrt{c}$.

4. Multiplied radical expressions. If a and b are greater than or equal to
    zero, then $\sqrt{ab} = \sqrt{a}\sqrt{b}$.

5. Divided radical expressions. If $a \geq 0$ and $b > 0$, then $\sqrt{\dfrac{a}{b}} = \dfrac{\sqrt{a}}{\sqrt{b}}$.

6. Rationalized square root expressions with monomial denominators;

7. Solved radical equations arising from the Pythagorean theorem;

8. Written radical expressions with fractional exponents and fractional
    exponents as radicals using $\sqrt[n]{a} = a^{\frac{1}{n}}$;

9. Solved radical equations graphically, numerically, and algebraically; and

10. Solved applications of radicals.

Unit 8 Problems for Practice

A. Review Problems. Follow the directions.

1. Complete the pattern for each.

a) 2, 5, 8, ___ , ___ , ___

b) 0, –4.5, –9, ___ , ___ , ___

2. Complete the I-O table. Plot the points. Is the graph linear?

x	$y = \dfrac{-2}{3}x + 4$
–3	
–1	
0	
1	
3	

3. Solve $2(-5t + 7) = t - (5t + 13)$ algebraically. Verify graphically or numerically.

4. Calculate the area and perimeter of the following figures.

2x – 3

2x – 3

8.2y

2 – 3y

5. 12 is what % of 120?

6. What is 30% of $59.95?

7. Solve by factoring:

a) $x^2 - 16 = 0$

b) $k^2 - 2k = 8$

8.   Solve the system of linear equations algebraically.

$$-2x + y = 6$$
$$x + 2y = -3$$

9.   Solve the system of linear equations graphically and verify algebraically.

$$3x - y = 7$$
$$x - 2y = -1$$

10.   Solve the system of quadratic equations numerically and confirm algebraically.

$$y = x + 2$$
$$x - y = -3$$

B.   Work the following review problems.

11.   Expand and simplify: $-5k(2n + 3r) + 4r(2n + 3r)$

12.   Divide and simplify: $\dfrac{40y^4 - 20y^3 + 10y^2}{10y^2}$

13.   The nucleus of a human liver cell is approximately 0.000004 meters in diameter. Write this in scientific notation.

14.   Given that the triangles are similar, use ratios to calculate the length of the unknown side.

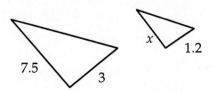

15. Under specific circumstances, the blood alcohol level of a person varies directly with the number of cans of beer consumed. If the person's blood alcohol level is 0.05% after drinking 2 cans of beer, what is the person's alcohol level after drinking 3 cans of beer?

16. Solve $5y - 2x + 20 = 0$ for y.

17. Factor $5x^2 - 180$.

18. Graph the line through $(-2, -5)$ and $(3, 8)$. Calculate the slope.

19. The chess club sold frozen pizzas to raise funds. Supreme pizzas sold for $5 each while sausage pizzas sold for $4 each. If the club raised $212 from the sale of 48 pizzas, how many of each type did they sell?

20. Approximate the solution for the system graphically, and determine the exact solution algebraically.

$$\begin{aligned} 2x - 3y &= 9 \\ 4x - 6y &= 12 \end{aligned}$$

C. Simplify the following:

21. $\sqrt{9}$

22. $\sqrt{121}$

23. $8\sqrt{100}$

24. $\sqrt{48}$

25. $\sqrt{45}$

26. $3\sqrt{700}$

27. $5\sqrt{72}$

28. $-3\sqrt{12}$

29. $-4\sqrt{196}$

30. $-\sqrt{28}$

D. Find the distance between the two points by plotting the points, drawing the right triangle, and using $d = \sqrt{a^2 + b^2}$.

31. (7, 3) and (5, 1)

32. (–1.4, 3.2) and (–1.4, 1.6)

33. (0, 4) and (–3, 0)

34. (–3, 1) and (–1, –1)

35. (10, 20) and (30, 40)

36. (37, 27) and (0, 0)

E. Simplify by combining like terms.

37. $3\sqrt{5} - 9 - 6\sqrt{5}$

38. $\sqrt{8} - \sqrt{12} + \sqrt{20}$

39. $\sqrt{18} - \sqrt{50} + \sqrt{12} - \sqrt{75}$

40. $\sqrt{2} - \left(-4\sqrt{8}\right)$

41. Find the perimeter of the rectangle with $l = \sqrt{75}$ and $w = \sqrt{12}$.

42. Find the perimeter of a parallelogram with $l = \sqrt{48}$ and $w = \sqrt{27}$.

43. Find the perimeter of a triangle with sides of length $\sqrt{72}$, $3\sqrt{3}$, and $\sqrt{49}$.

44. Find the perimeter of a rectangle with a length of $\sqrt{45}$ and a width of $\sqrt{20}$.

F. Simplify the following:

45. $\left(2\sqrt{3}\right)\left(\sqrt{6}\right)$

46. $\left(\sqrt{45}\right)\left(\sqrt{20}\right)$

47. $\left(3\sqrt{12}\right)\left(2\sqrt{3}\right)$

48. $\left(-4\sqrt{3}\right)\left(5\sqrt{15}\right)$

49. $\left(-5\sqrt{8}\right)\left(3\sqrt{12}\right)$

50. $2\sqrt{5}\left(\sqrt{5}-\sqrt{10}\right)$

51. $2\sqrt{5}\left(3\sqrt{5}-4\sqrt{7}\right)$

52. $-5\sqrt{14}\left(2\sqrt{7}-3\sqrt{2}\right)$

53. $-5\sqrt{8}\left(4\sqrt{3}-3\sqrt{6}\right)$

54. $\left(\sqrt{2}+\sqrt{3}\right)\left(\sqrt{2}-\sqrt{3}\right)$

55. $\left(\sqrt{2}+\sqrt{3}\right)\left(\sqrt{2}-\sqrt{5}\right)$

56. $\left(\sqrt{7}-\sqrt{5}\right)\left(\sqrt{3}-\sqrt{5}\right)$

57. $\left(2\sqrt{5}-3\sqrt{2}\right)\left(4\sqrt{10}-5\sqrt{2}\right)$

58. $\left(4\sqrt{6}-5\sqrt{3}\right)\left(2\sqrt{8}-5\sqrt{12}\right)$

59. If the length of a rectangle is $\sqrt{3}$ and the width is $\sqrt{12}$, what is the area?

60. What is the exact area of a triangle with a height of length $\sqrt{6}$ and base of length $\sqrt{2}$? What is the approximate area?

G. Divide and rationalize the denominator, if necessary.

61. $\dfrac{\sqrt{30}}{\sqrt{6}}$

62. $\dfrac{\sqrt{24}}{\sqrt{6}}$

63. $\dfrac{10\sqrt{48}}{4\sqrt{6}}$

64. $\dfrac{\sqrt{6}}{\sqrt{30}}$

65. $\dfrac{\sqrt{8}}{\sqrt{16}}$

66. $\dfrac{3\sqrt{3}}{5\sqrt{5}}$

67. $\dfrac{3\sqrt{8} - 2\sqrt{12}}{\sqrt{2}}$

68. $\dfrac{15\sqrt{30} - 20\sqrt{60}}{5\sqrt{15}}$

69. $\dfrac{-20\sqrt{20} - 30\sqrt{30}}{-5\sqrt{10}}$

70. $\dfrac{18\sqrt{18} - 36\sqrt{24}}{6\sqrt{6}}$

71. $\dfrac{\sqrt{3} - 2\sqrt{5}}{\sqrt{7}}$

72. $\dfrac{2\sqrt{7} - 3\sqrt{2}}{\sqrt{8}}$

73. A dropped object takes t seconds to fall a distance of d feet, where $t = \sqrt{\dfrac{d}{16}}$. Simplify this expression.

74. The period T of a simple pendulum is $T = \sqrt{\dfrac{4\pi^2 L}{g}}$, where L is the length of the pendulum and g is the acceleration due to gravity. Simplify this expression by rationalizing the denominator.

75. The Mach speed, M, of a jet can be determined from the formula $M = \sqrt{\dfrac{2}{\gamma}} \sqrt{\dfrac{P_2 - P_1}{P_1}}$, where P_1 and P_2 are air pressures and γ (gamma) is a constant. Simplify the right hand side of the formula.

76. The period T of a simple pendulum of length 2 feet is $T = \sqrt{\dfrac{8\pi^2}{32}}$. Simplify to give an exact answer and then approximate the period.

H. Complete the following table.

	Radical Expression	Exponential Expression	Approximation
77.	$\sqrt{8}$		
78.		$(60)^{\frac{1}{3}}$	
79.	$\sqrt[4]{33}$		
80.		$(8)^{\frac{1}{5}}$	
81.	$\sqrt[7]{33}$		
82.		$(800)^{\frac{1}{8}}$	
83.		$(8)^{-\frac{1}{5}}$	
84.	$\sqrt[5]{330}$		

I. Work each of the following problems.

85. Find the area of a triangle with sides of 4, 5.5, and 7.8 inches.

86. Find the third side of a right triangle if the hypotenuse is 66 cm and one side is 37 cm.

87. Find the hypotenuse of a right triangle with sides of 33 and 66 cm.

88. The escape velocity for a spacecraft is $v = \sqrt{2gR}$, where g is the planet's acceleration due to gravity and R is the radius of the planet. What is the escape velocity for a planet whose radius is 22,000 miles and $g = 30$ ft/s²? What is the escape velocity for earth? $g = 32$ ft/s² for the earth.

89. What is the period of a pendulum that is 33 inches long if g is 32 feet/sec²?

90. What is the period of a pendulum 33 centimeters long?

91. What is the velocity of a satellite orbiting at a distance of 16,967 km from the center of the earth?

92. What is the velocity of a satellite orbiting at a distance of 35,000 km from the center of earth?

93. What is the velocity of a satellite that is orbiting at a distance of 19,900 miles from the center of the earth?

94. The speed in miles per hour a car can travel on a highway curve without skidding is $v = \sqrt{9r}$, where r is the radius of the curve in feet. What should be the posted speed on a curve of 100 feet?

J. Solve each of the following algebraically and verify graphically or numerically.

95. $\sqrt{2x+3} = 5$

96. $\sqrt{x^2-3} = 7$

97. $\sqrt[4]{2x-5} = 3$

98. $\sqrt[3]{3-x^2} = 1$

99. How far from the center of the earth is a satellite orbiting at a velocity of 5000 meters per second?

100. The period of a pendulum is 1.5 seconds. How long is the pendulum? Use $g = 9.8 \text{ m/s}^2$.

Pretest of Prerequisite Skills

Perform the following calculations without use of a calculator. Choose the best answer and record it on the answer sheet.

1. 0.2 + 1.09

    a) 1.11
    b) 0.111
    c) 1.19
    d) 1.29

2. 9.08 − 6.324 =

    a) 3.764
    b) 2.764
    c) 2.756
    d) 2.656

3. 1.83×40.7

    a) 54.461
    b) 74.481
    c) 86.01
    d) 70.448

4. $32.4 \div 0.09 =$

    a) 360
    b) 36
    c) 3.6
    d) 0.36

5. Round 840.156 to the nearest hundredth.

    a) 840.16
    b) 800
    c) 840.15
    d) 900

6. $\dfrac{3}{8} + \dfrac{1}{4} =$

    a) $\dfrac{3}{32}$
    b) $\dfrac{4}{12}$
    c) $\dfrac{5}{12}$
    d) $\dfrac{5}{8}$

7. $\dfrac{5}{12} + \dfrac{3}{10} =$

    a) $\dfrac{15}{22}$
    b) $\dfrac{43}{60}$
    c) $\dfrac{15}{120}$
    d) $\dfrac{8}{22}$

8. $\dfrac{7}{9} - \dfrac{2}{3} =$

    a) $\dfrac{1}{9}$

    b) $\dfrac{1}{3}$

    c) $\dfrac{5}{27}$

    d) $\dfrac{5}{6}$

9. $7\dfrac{2}{3} - 4\dfrac{3}{4} =$

    a) $3\dfrac{1}{12}$

    b) $2\dfrac{1}{9}$

    c) $2\dfrac{11}{12}$

    d) $3\dfrac{11}{12}$

10. $\left(\dfrac{3}{4}\right)\left(\dfrac{4}{15}\right) =$

    a) $\dfrac{1}{5}$

    b) $\dfrac{1}{4}$

    c) $\dfrac{1}{3}$

    d) $\dfrac{1}{6}$

11. $\left(\dfrac{7}{8}\right)\left(\dfrac{40}{21}\right)$

    a) $\dfrac{3}{2}$

    b) $\dfrac{5}{3}$

    c) $\dfrac{147}{320}$

    d) $\dfrac{320}{147}$

12. $\dfrac{3}{4} \div \dfrac{6}{5} =$

    a) $\dfrac{9}{10}$

    b) $\dfrac{8}{5}$

    c) $\dfrac{4}{5}$

    d) $\dfrac{5}{8}$

13. $3\dfrac{1}{2} \div 2\dfrac{1}{3}$

    a) $1\dfrac{1}{6}$

    b) $1\dfrac{1}{2}$

    c) $\dfrac{2}{3}$

    d) $\dfrac{6}{7}$

14. $0.035 =$

    a)   35%
    b)   0.035%
    c)   0.0035%
    d)   3.5%

15. $\dfrac{4}{7} \approx$

    a)   0.57%
    b)   5.7%
    c)   57.1%
    d)   571%

16. $72\% =$

a) $\dfrac{1}{72}$

b) $\dfrac{39}{50}$

c) $\dfrac{18}{25}$

d) $\dfrac{19}{25}$

17. $250\% =$

a) 2.5

b) 0.25

c) 0.025

d) 25.0

18. 40% of $120 =$

a) 480
b) 48
c) 4.8
d) 0.48

19. 12 is what percent of 48?

a) 0.25%
b) 4%
c) 25%
d) 40%

20. $|-5|$

a) 5
b) 0
c) −5.5
d) 5.5

21. $-|-3| =$

a) 3
b) 0
c) −3
d) 4

22. $(-7) + (-12) =$

a) 19
b) −19
c) −5
d) 5

23. $(-3) + (+2) + (-7) =$

a) −8
b) 8
c) −2
d) 2

24. $(-3) - (-3) =$

a) −6
b) 0
c) 6
d) 9

25. $(-7) - (+3) =$

a) 10
b) −4
c) 4
d) −10

26. $-17 - 8 =$

    a) -25
    b) 25
    c) -9
    d) 9

27. $(+5)(-3) =$

    a) 2
    b) -2
    c) 15
    d) -15

28. $(-3)(+2)(-4) =$

    a) -5
    b) -24
    c) 24
    d) 5

29. $(-12) \div (-3) =$

    a) $\dfrac{1}{4}$
    b) $\dfrac{-1}{4}$
    c) -4
    d) 4

30. $-68 \div (-2) =$

    a) -32
    b) -34
    c) 34
    d) 36

31. $12 \div 4 + 2 =$

    a) 2
    b) 5
    c) 15
    d) -2

32. $5 - 6 \div 2 =$

    a) $\dfrac{-1}{2}$
    b) $\dfrac{1}{2}$
    c) 2
    d) -2

33. $(-3)^2 \cdot 5 + 4 =$

    a) -26
    b) 49
    c) 36
    d) -41

Answer Sheet

1. _____
2. _____
3. _____
4. _____
5. _____
6. _____
7. _____
8. _____
9. _____
10. _____
11. _____
12. _____
13. _____
14. _____
15. _____
16. _____
17. _____
18. _____
19. _____
20. _____
21. _____
22. _____
23. _____
24. _____
25. _____
26. _____
27. _____
28. _____
29. _____
30. _____
31. _____
32. _____
33. _____

Prerequisite Skills Exercises

Problems 1–10 keyed to Pretest 1–2

1. $5.34 + 6.7 + 11.001 =$

2. $6.7 + 18.23 =$

3. $9.01 + 12.389 =$

4. $12.25 + 198.38 =$

5. $10,398.27 + 2458.3 =$

6. $7.3 - 0.29 =$

7. $11.4 - 5.003 =$

8. $14,375.23 - 5875.88 =$

9. $6.23 - 2.48 =$

10. $121.67 - 73.57 =$

Problems 11–20 keyed to Pretest 3–4

11. $37.2 \times 6.4 =$

12. $4.2 \times 0.005 =$

13. $7.07 \times 8.97 =$

14. $(0.6)(0.0028) =$

15. $(0.854)(2.8) =$

16. $14.5 \div 0.05 =$

17. $4.5 \div 0.9 =$

18. $36.3 \div 3.63 =$

19. $5.4 \div 0.24 =$

20. $336 \div 0.24 =$

Problems 21–24 keyed to Pretest 5

21. Round 92.28 to the nearest tenth.

22. Round 2.28 to the nearest whole number.

23. Round 0.01459 to the nearest hundredth.

24. Round $52.399 to the nearest cent.

Problems 25–34 keyed to Pretest 6–9

25. $\dfrac{2}{3} + \dfrac{1}{6} =$

26. $\dfrac{1}{6} + \dfrac{3}{4} =$

27. $5\dfrac{2}{3} + 3\dfrac{1}{4} =$

28. $\dfrac{2}{5} - \dfrac{3}{10} =$

29. $\dfrac{2}{5} + \dfrac{3}{10} =$

30. $\dfrac{5}{12} + \dfrac{3}{8} =$

31. $\dfrac{2}{3} - \dfrac{1}{6} =$

32. $\dfrac{5}{12} - \dfrac{3}{8} =$

33. $\dfrac{3}{4} - \dfrac{1}{6} =$

34. $5\dfrac{2}{3} - 3\dfrac{1}{6} =$

Problems 35–44 keyed to Pretest 10–13

35. $\left(\dfrac{2}{3}\right)\left(\dfrac{5}{8}\right) =$

36. $\left(\dfrac{3}{7}\right)\left(\dfrac{21}{9}\right) =$

37. $\left(\dfrac{18}{55}\right)\left(\dfrac{10}{27}\right) =$

38. $\left(\dfrac{60}{27}\right)\left(\dfrac{81}{100}\right) =$

39. $\left(\dfrac{37}{144}\right)\left(\dfrac{72}{111}\right) =$

40. $\dfrac{21}{9} \div \dfrac{7}{3} =$

41. $\dfrac{8}{5} \div \dfrac{5}{2} =$

42. $\dfrac{27}{10} \div \dfrac{6}{5} =$

43. $\dfrac{100}{81} \div \dfrac{60}{27} =$

44. $2\dfrac{1}{2} \div 3\dfrac{1}{5} =$

Problems 45–54 keyed to Pretest 14–17

45. Change $\dfrac{4}{5}$ to a percent.

46. Change $\dfrac{3}{8}$ to a percent.

47. Change 1.33 to a percent.

48. Change 0.276 to a percent.

49. Change 22% to a fraction in simplest form.

50. Change 95% to a fraction in simplest form.

51. Change 72.5% to a decimal.

52. Change 278% to a decimal.

53. Change $33\dfrac{1}{3}\%$ to a fraction.

54. Change $11\dfrac{1}{9}\%$ to a fraction.

Problems 55–64 keyed to Pretest 18–19

55. 14% of 80 =

56. 212% of 11 =

57. 4 is what percent of 20?

58. 90 is what percent of 10?

59. 10% of what is 4?

60. 15% of what is 6?

61. 33,337.20 is what percent of $32,055?

62. 2.1% of $24,080 =

63. 6% of $1250 =

64. 96% of what is $31,609.92?

Problems 65–70 keyed to Pretest 20–21

65. $|-11| =$

66. $|45| =$

67. $-|-7| =$

68. $-|+3| =$

69. $|5-3| =$

70. $|3-5| =$

Problems 71–80 keyed to Pretest 22–26

71. $(-3)+(-7) =$

72. $(-11)+(+47) =$

73. $(-3)+(-5)+(-5) =$

74. $(-11)+(-5)+(-3) =$

75. $(-3)-(+5) =$

76. $-11-(-4) =$

77. $-2-(+3)-(-5) =$

78. $-5-(-4)-8 =$

79. $5-3-2+8-4 =$

80. $2-3+5-8-11 =$

Problems 81–88 keyed to Pretest 27–30

81. $(-3)(+7) =$

82. $(-5)(-4) =$

83. $(-3)^2 =$

84. $(-5)^3 =$

85. $(-2)(-3)(-5) =$

86. $(-2)(+5)(-7) =$

87. $(-20) \div (+4) =$

88. $\dfrac{-50}{-5} =$

Problems 89–100 keyed to Pretest 31–33

89. $\dfrac{30-50}{2-7} =$

90. $\dfrac{20-30}{7-2} =$

91. $5-3 \cdot 2 =$

92. $4+3 \cdot 5 =$

93. $-4(2-3)(5+4) =$

94. $7(3-5)(7-11) =$

95. $3^2 \cdot 4-5 =$

96. $5-3 \cdot 2^2 =$

97. $\left(-\dfrac{1}{3}\right)-\left(\dfrac{-1}{3}\right) =$

98. $10-4 \div 2 =$

99. $2^2-2 \cdot (12 \div 6) =$

100. $12-6+3 =$

Using the TI-83 Graphing Calculator

APPENDIX B

Contents

Checklist

These are some items to remember throughout the semester.

✓ Use the gray "(-)" for a negative sign and the blue "–" for subtraction.

✓ Check to see that you are in the **MODE** you want.

✓ Turn off your **STAT PLOTS** off before graphing a new function.

✓ Turn off all functions in the **Y**= screen before using **STAT PLOTS**.

✓ Use the caret key, ⌐∧⌐, for all exponents other than 2.

✓ Enclose all fractions in parentheses.

✓ Do not approximate fractions.

✓ Use your calculator efficiently.

✓ Remember the manual that came with your TI-83. It is a valuable resource.

✓ Have fun and explore the capabilities of your TI-83. You really can't break it by using it.

✓ Don't leave your TI-83 to bake in your car in the summer or freeze in the winter.

✓ Don't drop the TI-83. You can crack the display.

✓ Don't touch the screen of the TI-83. Touching the screen can damage it.

Using the TI-83 Graphing Calculator

In each case, an example will be followed by the appropriate keystrokes or instructions.

Finding the square of a number

Example: 4.2^2

Press $\boxed{4}$ $\boxed{.}$ $\boxed{2}$ $\boxed{x^2}$ $\boxed{\text{ENTER}}$. The correct answer is 17.64.
Remember: The square affects only the number immediately preceding it, so use parentheses to square a negative number.

Raising a number to a power

Example: 2.5^3

Press $\boxed{2}$ $\boxed{.}$ $\boxed{5}$ $\boxed{\wedge}$ $\boxed{3}$ $\boxed{\text{ENTER}}$. The $\boxed{\wedge}$ key is immediately above the $\boxed{\div}$ key. The correct answer is 15.625.

Finding the square root of a number

Example: $\sqrt{152.2}$

Press $\boxed{\text{2nd}}$ $\boxed{x^2}$ $\boxed{1}$ $\boxed{5}$ $\boxed{2}$ $\boxed{.}$ $\boxed{2}$ $\boxed{)}$ $\boxed{\text{ENTER}}$. The answer on the screen should be 12.33693641.

Finding the cube root of a number

Example: $\sqrt[3]{-27}$

Preferably, you would use the exponent $\frac{1}{3}$ to obtain the cube root of –27. The keystrokes would be $\boxed{(}$ $\boxed{(-)}$ $\boxed{2}$ $\boxed{7}$ $\boxed{)}$ $\boxed{\wedge}$ $\boxed{(}$ $\boxed{1}$ $\boxed{\div}$ $\boxed{3}$ $\boxed{)}$ $\boxed{\text{ENTER}}$.

Otherwise, press $\boxed{\text{MATH}}$ to show the screen at the right. Press $\boxed{4}$ for $^3\sqrt{\ }$ (. Then enter –27) and press $\boxed{\text{ENTER}}$. The correct answer is –3.

Multiplying by a fraction

Example: $\frac{9}{5} \cdot 10$

Press $\boxed{(}$ $\boxed{9}$ $\boxed{\div}$ $\boxed{5}$ $\boxed{)}$ $\boxed{\times}$ $\boxed{1}$ $\boxed{0}$ $\boxed{\text{ENTER}}$. The correct answer is 18. Observe that the fraction is enclosed by parentheses.

Corrections

You can use the arrow keys, $\boxed{\text{DEL}}$, and $\boxed{\text{2nd}}$ $\boxed{\text{DEL}}$ to correct work if you have not yet pressed $\boxed{\text{ENTER}}$. Once corrected, you can press $\boxed{\text{ENTER}}$ from any spot on that line. If you have already pressed $\boxed{\text{ENTER}}$, you can press $\boxed{\text{2nd}}$ $\boxed{\text{ENTER}}$ to retrieve previous lines of work.

Evaluating expressions with π

Example: $\frac{4}{3}\pi(3)^3$

Press [(] [4] [÷] [3] [)] [2nd] [^] [(] [3] [)] [^] [3] [ENTER]. The screen should match the screen to the right.

Plotting Points

Example: Plot the points in the following table:

x	1	2	3	4
y	12	24	36	48

Press [STAT] to get the screen to the right.

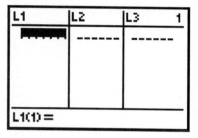

Press [1] or press [ENTER] so you can edit the data in the list.

You should have a screen that matches the one to the right. If not, look up clearing a list (page A-16) before proceeding. The **1** in the upper right-hand corner indicates the cursor is in the first column.

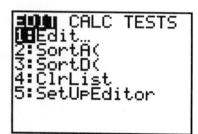

Now enter 1, 2, 3, and 4. Press [ENTER] after each number. To move to the next column press the right arrow [▶]. In the second column enter 12, 24, 36, and 48. Press [ENTER] after each number.

Now you are ready to plot the numbers. Press [2nd] [Y=] to access **STAT PLOT**. Your screen will look similar to this. Press [1] or [ENTER] to use **Plot1**.

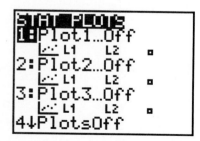

Tracing

Solve $3x - 9.6 = 0$ by tracing the graph of $y = 3x - 9.6$.

Graph $y = 3x - 9.6$ as described before. The solution of the equation is the x-value of the x-intercept. We can find the x-intercept by tracing the graph.

With the graph of $y = 3x - 9.6$ on the screen, press $\boxed{\text{TRACE}}$. Use the right and left arrows to trace the graph until you reach the point where the y-coordinate is 0 as shown. So the solution of $3x - 9.6 = 0$ is $x = 3.2$.

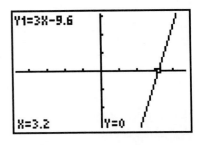

Using the TABLE feature

Example: If $h = p^2 + 2p + 3$, find h when $p = 85$ and $p = 100$.

Enter $x^2 + 2x + 3$ into $\boxed{\text{Y=}}$ using x instead of p and y for h. Press $\boxed{\text{2nd}}$ $\boxed{\text{WINDOW}}$ to access table setup (**TBLSET**).

Since we are interested in the point where $x = 85$, we might set the **TblStart** equal to 85. We also need to find the y-value when $x = 100$ so we could set the table change (Δ**Tbl**) equal to 5. (If we chose 10, we would miss $x = 100$. Why?) Leave **Indpnt** and **Depend** on **Auto**.

Finally, press $\boxed{\text{2nd}}$ $\boxed{\text{GRAPH}}$ for **TABLE**. The result should look like the screen at the right. So $h = 7398$ when $p = 85$, and $h = 10{,}203$ when $p = 100$.

Clearing a list

Example:

L1	**L2**	**L3**	**1**
1	12	------	
2	24		
3	36		
4	48		
------	------		

L1 ={1,2,3,4}

To remove the values from list **L1**, use the up arrow ▲ until the **L1** is backlit. Then press [CLEAR] and [ENTER]. Repeat this for any other lists you need to clear.

Evaluating an equation for one value

Example: If $y = 2x + 3.5$, what is y when x is 5.25?

Press [Y=], and enter $2x + 3.5$. Press [2nd][MODE] to return to the **HOME** screen. From this screen, press [VARS] (variables). Arrow over ▶ to highlight **Y-VARS** (y-variables). Choose option **1: Function...**, then option **1: Y1**. This process pastes **Y1** on whatever screen you started from; in this case, the home screen.

Now, enter [(] [5] [.] [2] [5] [)] and press [ENTER]. The resulting answer is 14 as shown here.

```
Y1(5.25)
                    14
```

Evaluating an equation for several values

Example: If $y = 2x + 3.5$, what is y when x is 2, 0, and 7.37?

Press [Y=], and enter $2x + 3.5$. Press [2nd] [MODE] to return to the **HOME** screen. From this screen, press [VARS] (variables). Arrow over ▶ to highlight **Y-VARS** (y-variables). Choose option **1: Function...**, then option **1: Y1**. This process pastes **Y1** on whatever screen you started from; in this case, the home screen.

Then, press [(] [2nd] [(] [2] [,] [0] [,] [7] [.] [3] [7] [2nd] [)] [)]. Press [ENTER]. Your screen should appear as the one at the right. When x is 2, y is 7.5. When x is 0, y is 3.5, and when x is 7.37, y is 18.24.

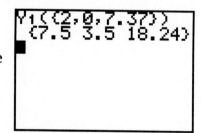

```
Y1({2,0,7.37})
      {7.5 3.5 18.24}
■
```

Graphing with shading

From the Y= screen, begin by pressing the left arrow so that the cursor is on the left hand side of the **Y1**=. Pressing ENTER changes the style of the graph. The cycle of styles is ╲ line, ▐ thick line, ▜ shade above, ▙ shade below, ⫯ path, ⫯ animate, and ⋰ dotted line. When it shows the style you want, press GRAPH.

The following screens are used to graph $y \geq 2x + 3.5$ using the graphing style that shades above the graph.

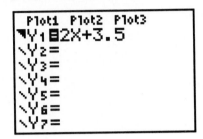
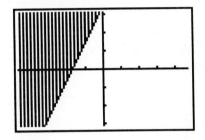

Using the table and graph simultaneously

Press MODE and change the last setting to **G-T**. If you have entered a function into the **Y=** menu, press GRAPH. Pressing TRACE allows you to observe the graph and table as you move across the graph. This feature only works with values that lie in the window on the screen.

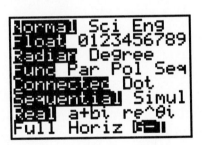
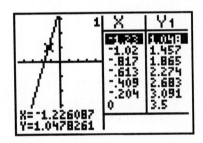

Troubleshooting

```
ERR:SYNTAX
1∎Quit
2:Goto
```

Your language does not match what the TI-83 expected. Press 2. Then closely examine what you told the TI-83 to do. For example, did you use "–" when you should have used the negative sign, "(-)"? Did you use parentheses correctly? You may choose to quit or go to the error location suggested by the calculator.

```
ERR:DIVIDE BY 0
1∎Quit
2:Goto
```

Division by zero is undefined; so look at your expression to see if an operation or variable makes the denominator of the fraction zero. This error message does not display during graphing even though there may be *x*-values that result in division by zero.

```
ERR:NONREAL ANS
1∎Quit
2:Goto
```

This error occurs when you have made a mathematical error in a computation. For example, did you try to take the square root of a negative number? This error does not occur during graphing.

```
ERR:DIM MISMATCH
1∎Quit
```

This error occurs when the number of entries in a list does not match the number of entries in another list being used to plot points. Check your lists for matching lengths. Check your **STAT PLOTS** to determine if two lists were unexpectedly paired.

```
ERR:INVALID DIM
1∎Quit
```

This error occurs when you are trying to graph a function, but you cleared your lists and forgot to turn **STAT PLOTS** off. **STAT PLOTS** can be turned off from the Y= menu. Just arrow up until **Plot1**, **Plot2** or **Plot3** is highlighted by the cursor and press ENTER.

Selected Answers

Comments: Throughout the answers, where not specified, approximate answers have been arbitrarily rounded to the nearest hundredth unless such rounding provides fewer than three significant digits.

Unit 1

1. a) 1019
    b) 6972
    c) ≈ -57.67
    d) ≈ 34.52
    e) -207
    f) 2925
3. a) 138
    b) 166
    c) 41
    d) $-27{,}240$
    e) 1119
5. a) 4.09
    b) 3.062
    c) 1.6
    d) 75.435
    e) 7.62
7. a) $4\frac{1}{8}$

    b) $2\frac{2}{9}$

    c) $\frac{7}{8}$

    d) $8\frac{103}{128}$

9. a) 0.1257
    b) 0.02987
    c) 3.247
    d) 2.30
    e) 0.7823
    f) 0.0201

11. a) 7
    b) \$30
    c) 19.6
    d) $\approx 3.11\%$
    e) 9%
    f) $\approx \$0.37$
13. a) 16 units^2
    b) 30 units^2
15. a) 36 units^2
    b) 48 units^2
17. $P = 42$ cm,
    $A = 108$ cm^2
19. a) $H = 5$ m
    b) $H \approx 7.21$ m
    c) $H \approx 10.77$ m
21. a) $S \approx 201.06$ cm^2
    b) $S \approx 132.73$ in^2
    c) $S \approx 706.86$ m^2
23. a) $r \approx 1.71\ \Omega$
    b) $r \approx 2.32\ \Omega$
    c) $r \approx 2.48\ \Omega$
25. 7.28, 8.32, 9.22
27.

P	A
8 in.	4 in^2
18 cm	20.25 cm^2
25 1/3 ft	40 1/9 ft^2
30 m	56.25 m^2
400 yd	10,000 yd^2
1333.32 km	111,108.89 km^2

29. $0.8\ \Omega$, $1.22\ \Omega$, $1.87\ \Omega$,
    $2\ \Omega$, $2.74\ \Omega$, $3.85\ \Omega$
31. 0, 75, 380, 225, 150, 100

Unit 1, continued

33. **a)** −260 trillion
    b) −40 trillion
    c) −910 trillion
35. −9, −6, −3
37. 0, −25, −50
39. 64, 125, 216
    perfect cubes
41. $225, $250
43. −1, 1, 3, 5, 7
    linear

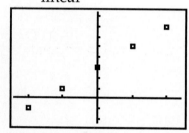

45. 7, 4, 3, 4, 7

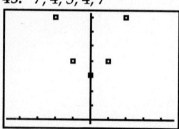

47. 0, 1.41, 2, 2.45, 3.46

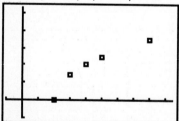

Unit 2

1. $\dfrac{5}{12}$
3. $A = 21$ units^2, $P = 20$ units
5. 67.5 cm^2

7. 96, 93.5, 91, 88.5, 86, 83.5
9. **a)** −16, −20, −24
    b) 13, 16, 19
11.

Year	$$
1994	
1996	
	19,400
	20,000
	19,000

13. equation
    exp1 = $2x - 5$
    exp2 = 4
15. expression
17. equation
    exp1 = $4(2x + 3)$
    exp2 = $3 - (2x - 7)$
19. equation
    exp1 = $5x^2$
    exp2 = 80
21. $x = 3$
    Tables may vary.

x	$3x+1$
1	4
2	7
3	10
4	13
5	16

23. $b = -3.5$
25. $x = -1$
27. $x = 2$
29. $d = 9.5$
31. $g = 12.2$
33. **a)** $x = 3$
    b) $x = 2$
    c) $x = 1$
    d) $x = -1$

Unit 2, continued

35. a) $x = 7\frac{1}{2}$

b) $x = 3$

c) $x = -1\frac{1}{2}$

d) $x = -6$

37. a) $x = -1.4, x = 1.4$

b) $x = -2, x = 2$

c) $x = -3, x = 3$

d) $x = -3.46, x = 3.46$

39. $x = 4.5$

x	3	3.5	4	4.5	5
y	1	2	3	4	5

41. $x = 24$

x	12	15	18	21	24
y	3	5	7	9	11

Windows for 43 and 45 will vary.

43. **Y1** $= -2x - 5$,   **Y2** $= 3$
window: $[-4.7, 4.7]_x$ and $[-3, 5]_y$
$x = -4$

45. **Y1** $= 45x - 17$, **Y2** $= 532$
window: $[-1, 18.8]_x$ and $[-5, 550]_y$
$x = 12.2$

47. $x = \frac{2}{3}$

49. $z = \frac{3}{2}$

51. $x = -5.5$

53. $a = -1.75$

55. $b = 7\frac{1}{5} = 7.2$

57. $6x - 8$

59. $6 + 15y$

61. $-15a + 20$

63. $-42 + 12b$

65. $-10 + 6x$

67. $x = \frac{7}{13}$

69. $x = -\frac{18}{5}$

71. $x = -1$

73. $x = 15.5$

75. $x = -2.5$

77. Width is 10 cm.

79. $A = (x + 4)(2x + 4) - 2x^2$

Unit 3

1. $x = \frac{23}{3} \approx 7.67$

3. $y = 12$

5. $x = \frac{17}{5} = 3.4$

7. $c \approx 0.20$

9. She needs to use 4 pounds of meat.

11. $x = 11.5$

13. $y = -8$

15. $x \approx 2.43$

17. $c \approx -0.53$

19. The maximum allowed speedometer reading is approximately 137 kph.

21.

Speed	Distance
	115 ft
	200 ft
	230 ft
80 mph	
30 mph	

23.

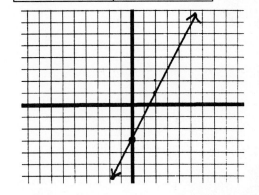

Unit 3, continued

25.

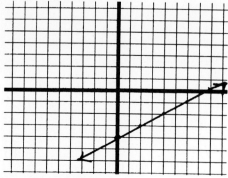

27.

t, yr.	L, mm
0	5.18
1	5.18
2	5.18
3	5.18
4	5.19
5	5.19
6	5.19

29.

h, in.	W, pounds
66	140
67	145
68	150
69	155
70	160
71	165
72	170
73	175
74	180

31. $P = 4x + 6y$

33. $P = 7z + w$

35. $P = 2b + 3a + 10$

37. $P = 10x$

39. $11x + 1$

41. $-3t - 8s$

43. $2x^2 + 3x + 1$

45. $3c^3 + 3c - 5$

47. $18x - 23$

49. $-2x + 10y + 15$

51. $10ab$

53. $21h + 3hw$

55. $\dfrac{7}{2}ab + 7a$

57. $6x^2 + xy - y^2$

59. $10x^2 + 15x$

61. $6c^2 - 4cd - 2d^2$

63. $6x^2 - 23x + 21$

65. $25 - 9v^2$

67. $6x^2 - 8xy - 5x - 56 + 28y$

69. $3x^5$

71. $-15x^7y^5$

73. $-12a^7b^5$

75. $35r^{10}t^7z^6$

77. $\dfrac{15a^2}{b^3}$

79. $x + \dfrac{y}{x}$

81. $\dfrac{1}{9x^2}$

83. 1

85. $-\dfrac{1}{64a^9}$

87. $5x^2y + 5y^5$

89. $\dfrac{10.8b}{a^6} - 9ab$

91. $\dfrac{p^2}{q^3} + \dfrac{p}{q^2} + \dfrac{p^2}{q^6}$

93. $\dfrac{x}{y} + \dfrac{11y^8z^{23}}{x^9}$

95. $-5x^3$

97. $x - 4$

99. $3a^3 - \dfrac{4c^3}{a^2}$

101. $\dfrac{w^2}{z} + 5w$

103. $5p^2 + 2p - 1$

105. $4x^4 - 6x^2 + \dfrac{3}{x}$

107. 9.35×10^{-4}

109. 6.642×10^5

Unit 3, continued

111. 0.000235

113. 5,780,000

115. 1.62×10^{13}

117. 7×10^{14}

119. 4.84×10^{20}

121. **a)** $(8 \times 10^{-8})(3 \times 10^{-8})$

    b) 2.4×10^{-15}

123. **a)** $(3 \times 10^{12})^2$

    b) 9×10^{24}

125. **a)** $(4 \times 10^{10})^{-3}$

    b) 1.6×10^{-32}

127. **a)** 3×10^{-11}

    b) 3.10×10^{-11}

129. **a)** 2.7×10^{37}

    b) 2.76×10^{37}

131. **a)** 6×10^{-19}

    b) 6.13×10^{-19}

133. **a)** 2.96×10^{28}

    b) 3.12×10^{28}

135. 4×10^{-6} m

137. 2.88×10^7 km

139. 7.7×10^9 bytes

141. $\dfrac{3 \times 10^6 \text{ pounds}}{4.5 \times 10^6 \text{ dollars}}$

    6.67×10^{-1} pounds/dollar

143. **a)** 5.4×10^5 bites/hour

    b) 2.07×10^{-2} L

    c) 2.484 L is a little less than half of 5.5 L

Unit 4

1. $x = \dfrac{-23}{5} = -4.6$

3. $y = \dfrac{102}{11} \approx 9.27$

5. $x = 6.8$

7. $c = 0.625$

9. $x + x + 7 = 163$

    85 ounces or 5 pounds 5 oz.

11. $P = 2L + 2W + 10$ units

    $A = 5L + LW$ units^2

13. $C = 10\pi x^2$ units

    $A = 25\pi x^4$ units^2

15. $P = 4s + 12$ units

    $A = s^2 + 6s + 9$ units^2

17. $P = 8h + 3b + b^2 + 3.5$ units

    $A = \dfrac{b^2 h}{2} + b^2 + 0.75h + 1.5$ units^2

19. $P = 6a\pi$ units

    $A = 9a^2\pi$ units^2

21. $16t + 5s$

23. $18x - 23$

25. $24x - 36y + 215$

27. $-9a^2 + 27a - 20$

29. $4a^2 - 25b^2$

31. $-6x^5$

33. $\dfrac{-15}{x^7 y^3}$

35. $\dfrac{1}{125x^3}$

37. 1

39. $5x^3 y + 5xy^5$

41. $\dfrac{p^3}{q^3} - \dfrac{p^2}{q^2} + \dfrac{p^3}{q^2}$

43. $2x - 1$

45. $4r^4 - 6r^2 + \dfrac{3}{r}$

47. Estimate 1.6×10^{17}

    Answer 1.69×10^{17}

49. Estimate 1.3×10^{21}

    Answer 1.34×10^{21}

51. 6×10^8

53. 1.003×10^{11} km

55. 2.68×10^8 people

57. $x = 13.8$

59. $r = 7.5$

61. $x \approx 3.49$

63. $R = \dfrac{2}{3}$ ohms, $R_3 = 7.5$ ohms

65. 2500 units

Unit 4

67. $x = \dfrac{35}{3} \approx 11.67$

69. $q = \dfrac{-47}{5} = -9.4$

71. 825 miles

73. 2.8×10^7 bytes, 72 disks

75. 2×10^{10} pounds

77. 4.5 hours

79. ≈ 73.3 mph

81. ≈ 2.6 ft

83. ≈ 53.57

85. 215

87. about 53.2%

89. $\approx \$9887$

91. $53.66

93. $y = \dfrac{5x}{6} - 5$

95. $y = \dfrac{-2}{3}x + 4$

97. $y = mx + 2m + 3$

99. $B = \dfrac{2A}{H}$

Base
12 in.
10 ft.
16 m
≈ 34 yd.

101. $l = \dfrac{1}{2}p - w$

l
12.5 cm
5.5 m
5.5 mm
97.65 m

103. $x < \dfrac{11}{3}$

$\left(-\infty, \dfrac{11}{3}\right)$

105. $x > 1.8$

$(1.8, \infty)$

107. $\dfrac{-1}{2} < x < 5$

$\left(\dfrac{-1}{2}, 5\right)$

109. a) $20 < T < 40$

b) $-6.67 < T < 4.44$

111. $C \leq \$110{,}000$

113.

115.

117.

Unit 5

1. $b = -3$

3. $x = \dfrac{22}{9} \approx 2.44$

5. $x = 26$

7. $k = 20$

9. about 15 cups

11. about 13 servings

13. 5.3 ounces

15. $x = 2.8$

17. $-4a^2 - 19a + 15$

19. $a^2 + \dfrac{a^3}{b} - a^3 b$

21. $x \approx -13.67$

23. $z = \dfrac{7}{2}y - \dfrac{11}{2}$

25. 157.5 cm

27. $39.60

29. 4752 yd.; ≈ 2.84 miles

Unit 5, continued

31. $x < 2$

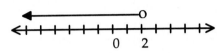

33. $p \le \dfrac{-12}{5}$

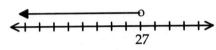

35. $x < 27$

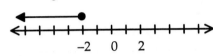

37. $\dfrac{5}{2} \le x < 6$

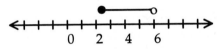

39. 130 mi. $\le x \le$ 145 mi.

41. $135.71 $\le x \le$ $178.57

43. 15 students

45. $m = 5$ $(0, 0)$
   $y = 5x$

47. $m = 3$ $(0, -3)$
   $y = 3x - 3$

49. $m = -5$ $(0, 0)$
   $y = -5x$

51. $m = 0$ $(0, -5)$
   $y = -5$

53. $m = \dfrac{1}{2}$ $(0, 3)$

   $y = \dfrac{1}{2}x + 3$

55. $m = -0.6$
   $y = -0.6x - 2$

57. $m = -6$
   $y = -6x + 40$

59. $m = \dfrac{-2}{5}$

   $y = \dfrac{-2}{5}x + \dfrac{7}{5}$

61. $m = -\dfrac{1}{60}$

   $y = \dfrac{-1}{60}x - \dfrac{47}{30}$

63. yes, linear
   $m = 4$
   $y = 4x - 12$

65. $90/hour

67. 80,650 persons/year

69.

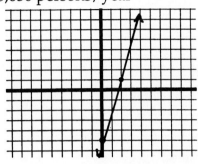

71.

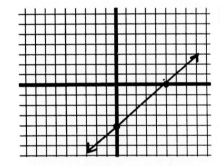

73.

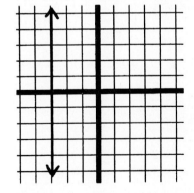

75.

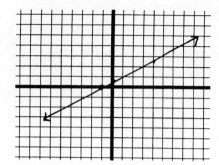

77.

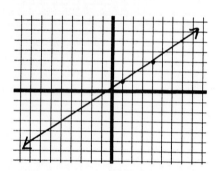

79.

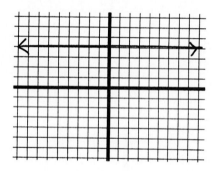

81.

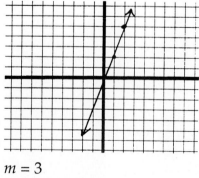

$m = 3$

83.

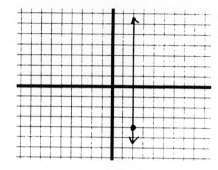

85.

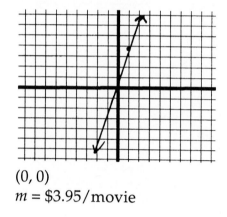

$(0, 0)$

$m = \$3.95/\text{movie}$

87.

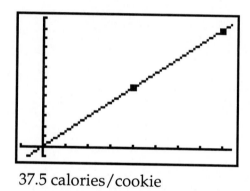

37.5 calories/cookie

Unit 6

1. $x \approx 14.82$
3. 3.75
5. $Q = 216$
7. 6.8 ft/sec, approximately
9. 4.29
11. 3
13. 82 mph, approximately
15. $y = -\dfrac{2}{3}x + 4$
17. $k = \dfrac{y}{x}$
19. $T = \dfrac{PV}{Nk}$
21. $y = 2x - 6$

x	y
–5	–16
–3	–12
0	–6
3	0
5	4

23. $m = \dfrac{-2}{3}$; y-intercept: $(0, 2)$

   x-intercept: $(3, 0)$

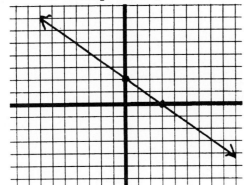

25. slope is undefined
    y-intercept: none
    x-intercept: $(-4, 0)$

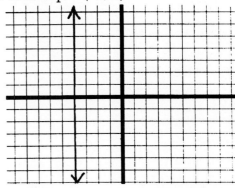

27. $y = \dfrac{2}{3}x - 3$
29. $y = \dfrac{7}{5}x + 2$
31. $x(x + 2)$
33. $4xy(xy + 2)$
35. $(x + 5)(x + 1)$
37. $(x - 4)(x + 2)$
39. $(x - 2)(x + 10)$
41. $(x - 6)(x + 6)$
43. $4(x - 3)(x + 2)$
45. prime
47. $(2 - x)(4 + x)$
49. $t(t - 3)(t + 2)$
51. $x = 0$ or $x = 2$
53. $x = -2$ or $x = 4$
55. $x = -10$ or $x = 2$
57. $x = -6$ or $x = 6$
59. $x = -2$ or $x = 3$
61. $v = -5$, $v = -1$, or $v = 0$
63. 534 feet
65. 3 inches by 14 inches
67. 4 feet by 12 feet
69. $x = -3.4$ or $x = 2.3$
71. $x = -0.5$ or $x = 4$
73. $x \approx -1.45$ or $x \approx 3.45$
75. 14%, approximately
77. 6 inches by 8 inches
79. $(x + 3)(x + 4)$
81. $6(x - 0.67)(x + 1.5)$
83. prime

85. $\dfrac{33+91a}{21a^2}$

87. $\dfrac{9+20t}{15t^3}$

89. $\dfrac{4z^2}{3}$

91. $\dfrac{(x-1)(x-7)}{2x}$

93. $P = \dfrac{B+A}{AB}$

95. $c = \dfrac{5}{6}$

97. no solution

99. 19 mi/hr, approximately

101. 47

103. $x = 7$ is the only reasonable value for x. The lengths of the sides are 3 cm and 3 m.

Unit 7

1. 78.54 sq. cm

3.

-2	-1
-1	-3
0	-5
1	-7
2	-9

yes, linear

5. $A = 6\,a^2$
$P = 12a$

7. 20%

9. $p = 4$ or $p = -2$

11. $b = -3.5$

13. $x = 0.4$

15. 1.475×10^{11} km

17. $v = -3$ or $v = 7$

19. $m = \dfrac{3}{5}$

21. $(3, 6)$

23. $(3, 1)$

25. $(2, -2)$

27. $(-1, 4)$

29. $\left(\dfrac{5}{4}, \dfrac{5}{6}\right)$ or about $(1.25, 0.83)$

31. no solution, lines are parallel

33. infinite number of solutions, lines coincide

35. 17 pecan, 28 chocolate

37. 10 caps

39. 200/year or about 16/month

41. $\left(\dfrac{20}{11}, \dfrac{12}{11}\right)$

43. $(3, 6)$

45. $(-1, 4)$

47. $(-6.5, 6)$

49. no solution, lines are parallel

51. no solution, lines are parallel

53. 45 cookies, 60 brownies

55. 11 bags

57. 66.5 minutes

Unit 8

1. 11, 14, 17
−13.5, −18, −22.5

3. $t = \dfrac{27}{6} = 4.5$

5. 10%

7. **a)** $x = -4$ or $x = 4$
b) $k = -2$ or $k = 4$

9. $(3, 2)$

11. $-10kn - 15kr + 8rn + 12r^2$

13. 4.0×10^{-6} meters

15. 0.075%

17. $5(x - 6)(x + 6)$

19. 20 supreme and 28 sausage

21. 3

23. 80

25. $3\sqrt{5}$

27. $30\sqrt{2}$

29. −56

31. ≈2.83

33. 5

35. ≈28.28

Unit 8, continued

37. $-3\sqrt{5} - 9$

39. $-2\sqrt{2} - 3\sqrt{3}$

41. $14\sqrt{3}$

43. $6\sqrt{2} + 3\sqrt{3} + 7$

45. $6\sqrt{2}$

47. 36

49. $-60\sqrt{6}$

51. $30 - 8\sqrt{35}$

53. $-40\sqrt{6} + 60\sqrt{3}$

55. $-\sqrt{15} - \sqrt{10} + \sqrt{6} + 2$

57. $40\sqrt{2} - 24\sqrt{5} - 10\sqrt{10} + 30$

59. 6

61. $\sqrt{5}$

63. $5\sqrt{2}$

65. $\dfrac{\sqrt{2}}{2}$

67. $6 - 2\sqrt{6}$

69. $6\sqrt{3} + 4\sqrt{2}$

71. $\dfrac{\sqrt{21}}{7} - \dfrac{2\sqrt{35}}{7}$

73. $t = \dfrac{\sqrt{d}}{4}$

75. $M = \dfrac{\sqrt{2\gamma P_1 (P_2 - P_1)}}{\gamma P_1}$

	Radical Expression	Exponential Expression	Approx.
77.	$\sqrt{8}$	$(8)^{\frac{1}{2}}$	2.83
79.	$\sqrt[4]{33}$	$(33)^{\frac{1}{4}}$	2.40
81.	$\sqrt[7]{33}$	$(33)^{\frac{1}{7}}$	1.65
83.	$\dfrac{1}{\sqrt[5]{8}}$	$(8)^{-\frac{1}{5}}$	0.66

85. 10.38 in^2

87. 73.79 cm

89. 1.84 sec

91. 4855 m/sec

93. 3534 m/sec

95. 11

97. 43

99. 1.6×10^7 m/sec

Prealgebra Skills

1. 23.041

3. 21.399

5. 12,856.57

7. 6.397

9. 3.75

11. 238.08

13. 63.4179

15. 2.3912

17. 5

19. 22.5

21. 92.3

23. 0.01

25. $\dfrac{5}{6}$

27. $8\dfrac{11}{12}$

29. $\dfrac{7}{10}$

31. $\dfrac{1}{2}$

33. $\dfrac{7}{12}$

35. $\dfrac{5}{12}$

37. $\dfrac{4}{33}$

39. $\dfrac{1}{6}$

41. $\dfrac{16}{25}$

43. $\dfrac{5}{9}$

45. 80%

Prealgebra Skills, continued

47. 133%

49. $\dfrac{11}{50}$

51. 0.725

53. $\dfrac{1}{3}$

55. 11.2
57. 20%
59. 40
61. 104%
63. $75
65. 11
67. −7
69. 2
71. −10
73. −13
75. −8
77. 0
79. 4
81. −21
83. 9
85. −30
87. −5
89. 4
91. −1
93. 36
95. 31
97. 0
99. 0

Index

Student Questionnaire

Elementary Algebra: A Prerequisite for Functions,
Preliminary Edition

Your comments on this edition will be very helpful to the authors and to Addison Wesley Longman as we work to make this the best possible textbook. Please take a few moments to complete this questionnaire and return it to Addison Wesley Longman. Postage is prepaid on the reverse side.

Name (optional): _____

School: _____

Course (name and number): _____

1. Did you cover all the units in class? If not, which units did you skip? _____

2. How satisfied were you with each of the following features:

	Not Satisfied	Satisfied	Very Satisfied
Unit objectives			
Collaborative problems or activities (those marked with the lizard icon)			
Group tips (also marked with the lizard icon)			
Other student interactions (those marked with the pencil-and-paper icon)			
Examples (those marked with the magnifying glass icon)			
You Try It exercises			
Unit projects			
Unit summaries			
Unit practice problems			
Appendix A: *Pretest of Prerequisite Skills*			
Appendix B: *Using the TI-83 Graphing Calculator*			
Appendix C: *Selected Answers*			
Index			

If you marked "Not Satisfied" for any of the above, please explain why this feature was unsatisfactory.

_____

_____

3. What did you like best about this textbook? _____

4. What did you like least about this textbook? _____

5. If you could change anything about this textbook, what would you change? _____

6. What unit was most difficult for you, and why? _____

7. Overall, was the textbook easy for you to read? Why or why not? _____

_____

_____

NO POSTAGE
NECESSARY
IF MAILED
IN THE
UNITED STATES

BUSINESS REPLY MAIL
FIRST CLASS   PERMIT NO. 11   READING, MA

Postage Will Be Paid By Addressee

Addison Wesley Longman
Attn: Laura Rogers
One Jacob Way
Reading, MA 01867-9903

8. Did you purchase the *Student Support Manual* that accompanies this textbook? If so, was it helpful?

_____

_____

9. Is there anything that you would add to the *Student Support Manual* to make it a better learning tool?

_____

_____

10. Did you use the tutorial software that accompanies this textbook? If so, was it helpful?

_____

_____

11. Would you recommend this course and textbook to a friend? Why or why not?

_____

_____